Christoph Mang

Magnetic study of rocks from the Chesapeake Bay impact structure, USA

Christoph Mang

Magnetic study of rocks from the Chesapeake Bay impact structure, USA

Südwestdeutscher Verlag für Hochschulschriften

Impressum / Imprint

Bibliografische Information der Deutschen Nationalbibliothek: Die Deutsche Nationalbibliothek verzeichnet diese Publikation in der Deutschen Nationalbibliografie; detaillierte bibliografische Daten sind im Internet über http://dnb.d-nb.de abrufbar.

Alle in diesem Buch genannten Marken und Produktnamen unterliegen warenzeichen-, marken- oder patentrechtlichem Schutz bzw. sind Warenzeichen oder eingetragene Warenzeichen der jeweiligen Inhaber. Die Wiedergabe von Marken, Produktnamen, Gebrauchsnamen, Handelsnamen, Warenbezeichnungen u.s.w. in diesem Werk berechtigt auch ohne besondere Kennzeichnung nicht zu der Annahme, dass solche Namen im Sinne der Warenzeichen- und Markenschutzgesetzgebung als frei zu betrachten wären und daher von jedermann benutzt werden dürften.

Bibliographic information published by the Deutsche Nationalbibliothek: The Deutsche Nationalbibliothek lists this publication in the Deutsche Nationalbibliografie; detailed bibliographic data are available in the Internet at http://dnb.d-nb.de.

Any brand names and product names mentioned in this book are subject to trademark, brand or patent protection and are trademarks or registered trademarks of their respective holders. The use of brand names, product names, common names, trade names, product descriptions etc. even without a particular marking in this works is in no way to be construed to mean that such names may be regarded as unrestricted in respect of trademark and brand protection legislation and could thus be used by anyone.

Coverbild / Cover image: www.ingimage.com

Verlag / Publisher:
Südwestdeutscher Verlag für Hochschulschriften
ist ein Imprint der / is a trademark of
OmniScriptum GmbH & Co. KG
Heinrich-Böcking-Str. 6-8, 66121 Saarbrücken, Deutschland / Germany
Email: info@svh-verlag.de

Herstellung: siehe letzte Seite /
Printed at: see last page
ISBN: 978-3-8381-3734-6

Zugl. / Approved by: Karlsruhe, Karlsruher Institut für Technologie, Diss, 2012

Copyright © 2013 OmniScriptum GmbH & Co. KG
Alle Rechte vorbehalten. / All rights reserved. Saarbrücken 2013

Abstract

This thesis is a rock magnetic and mineral magnetism study of lithological units associated with an irregular magnetic field anomaly over the Chesapeake Bay impact structure (CBIS), Virgina, USA. It contains a profound characterization of shock- and post-shock-related microstructural features occurring in magnetite (Fe_3O_4) and pyrrhotite (Fe_7S_8). Both minerals mainly control the rock magnetic properties of the units related with a positive anomaly pattern over the CBIS. In a second step, the observations from the microstructural investigations are combined with those obtained from high- and low-temperature magnetic measurements. The results of this study provide a contribution to a general understanding of the magnetic properties of shocked rocks. Furthermore, they help to interpret impact-related magnetic field anomalies.

The magnetic field anomaly pattern associated with the CBIS shows various circular short-wavelength variations occurring within and outside the crater structure and these variations indicate specific magnetic sources in the crater fill. Examinations of borehole samples from four different drilling locations (Eyreville and Cape Charles in the central crater, Bayside and Langley in the annular trough) imply that only basement and suevite rocks carry magnetizations high enough to produce such anomalies. Among these units only a schist megablock and the suevite layer carry a remanent magnetization. The main magnetic carriers in all units are magnetite and pyrrhotite. Magnetite is subdivided into three types, named primary, shocked and secondary magnetite. Whereas primary magnetite is mainly characterized by grain sizes lying in the multidomain (MD) range, shocked magnetite has been largely affected by shock-induced brittle deformation that resulted in grain sizes producing pseudo-single-domain (PSD) behaviour. Shocked magnetite occurs within the suevite and in basement-derived gneiss blocks. The grains show two sets of planar fractures (PFs); they are strongly altered and some are partially molten. Secondary magnetite formed after the impact within the suevite and is present in form of clusters. Such a cluster consists of various needle-shaped grains that range from few nm up to ~10 μm in size. The small-sized grain fraction in these clusters is superparamagnetic (SP) and has lower Curie temperatures (T_C: 520 - 560°C) than single domain (SD) and MD magnetite (T_C: 580°C). Large fractions of shocked and secondary magnetite show a reduced or suppressed Verwey transition (T_V) and this observation can be linked with a non-stoichiometric composition of these magnetites. Non-stoichiometry is mainly a product of oxidation, which first affects the grain rims and subsequently incorporates larger volume fractions. The ratio between the oxidized and stoichiometric volume fraction determines if and how strong T_V is modified.

Besides Magnetite, the suevite contains shocked pyrrhotite, and a large fraction of these grains is strongly depleted in iron resulting in a smythite (Fe_9S_{11})-like composition. This pyrrhotite, referred to as iron-deficient pyrrhotite, shows an absent or a largely suppressed transition at 34 K and the T_C is higher (340 - 365°C) than for 4C pyrrhotite (Fe_7S_8: 320°C). Besides fracturing and large amounts of stacking faults, this mineral shows no significant microstructural deformation features if compared to experimentally shocked pyrrhotite. Since stacking faults are also a typical feature described for smythite, a reasonable conclusion is that alteration has entirely modified shocked pyrrhotite. Iron loss and an establishment of a magnetically ordered structure, which is different to that of 4C pyrrhotite, can be therefore ascribed to alteration and associated diffusion processes. Since the stacking faults cause large disordering of the crystal structure, ferrimagnetism presumably arises from defect-free domains of about 10 nm in diameter.

In a first stage of this thesis the deformation-like microstructures in iron-deficient pyrrhotite were interpreted to have been caused by the shock wave and it was difficult to separate shock- from alteration-induced features. However, shock experiments at pressures between 3 and 30 GPa on natural pyrrhotite ore revealed that shock-related microstructures in pyrrhotite, which have not been modified by subsequent alteration, look different to those observed in the iron-deficient pyrrhotite. Pyrrhotite shocked up to 8 GPa is characterized by mechanical deformation, including grain size reduction and the formation of PFs, planar deformation features (PDFs), and abundant internal defects. Associated with these features is an increase in coercivity and saturation isothermal remanent magnetization (SIRM), indicating an increase in SD behaviour. Shock pressures between 20 and 30 GPa produce large amorphous domains in the crystal lattice and at 30 GPa shock-melting occurs, which is associated with the formation of native iron.

The results of this study show that shock induces a series of structural and magnetic modifications that are additionally modified by subsequent alteration in a natural environment. Rock magnetic properties of impact crater units are therefore dependent on the pre-impact condition of the particular magnetic carriers and their impact-related deformation and alteration history. Within the impact sediments the non-uniform spatial distribution of magnetic minerals is another important factor, since locally enhanced amounts of magnetic minerals increases the bulk magnetization in this area. If a pronounced thermoremanent magnetization related to impact melt is missing in an impact structure, the magnetic field is not necessarily related to the crater structure, but is rather resulting from a complex interplay between various impact-related processes occurring from nm- to km-scale.

Zusammenfassung

Die vorliegende Arbeit ist eine gesteinsmagnetische und magneto-mineralogische Studie über Gesteinseinheiten der Chesapeake Bay Impakt Struktur (CBIS) in Virginia, USA. Einige der untersuchten Gesteine sind maßgeblich für diverse positive Anomalien des dortigen Magnetfeldes verantwortlich. Die wichtigsten magnetischen Träger dieser Einheiten sind dabei Magnetit (Fe_3O_4) und Pyrrhotin (Fe_7S_8). Im folgenden werden die schock- und alterationsbedingten Mikrostrukturen dieser beide Minerale detailliert charakterisiert und mit den Ergebnissen aus magnetischen Hoch- und Tieftemperaturmessungen in Zusammenhang gebracht. Die in dieser Arbeit präsentierten Beobachtungen und Interpretationen leisten einen Beitrag zum allgemeinen Verständnis von Impakt-bezogenen Prozessen und deren Auswirkungen auf die gesteinsmagnetischen Eigenschaften betroffener Lithologien.

Außerhalb und im Zentrum des magnetischen Anomalienmusters der CBIS treten einige zirkulare, kurzwellige Anomalien auf, die auf magnetische Quellen innerhalb der Kraterfüllung hindeuten. Untersuchungen an Proben von vier verschiedenen Bohrungen (Eyreville und Cape Charles in Kraterzentrum, Bayside und Langley im äußeren Krater) zeigen, dass nur Gesteine aus dem Grundgebirge und eine Suevitlage als solche Quellen in Frage kommen. Von diesen Einheiten tragen jedoch nur ein Schiefer-Megablock und der Suevit eine remanente Magnetisierung. Im Fokus dieser Arbeit stehen die beiden magnetischen Minerale Magnetit und Pyrrhotin, da diese im Wesentlichen die gesteinsmagnetischen Eigenschaften der untersuchten Lithologien bestimmen. Magnetit wird in dieser Studie in drei Untergruppen unterteilt: primärer, geschockter und sekundärer Magnetit. Während sich primärer Magnetit überwiegend durch Korngrößen im Multidomänen (MD) Bereich auszeichnet, wurde geschockter Magnetit durch spröde Deformation, hervorgerufen durch die Schockwelle, in kleinere Körner zerbrochen. Letztere zeigen überwiegend ein Pseudo-Single-Domänen (PSD)-Verhalten. Geschockter Magnetit tritt im Suevit und in aus dem Grundgebirge stammenden Gneisblöcken auf. Die einzelnen Körner weisen bis zu zwei Richtungen an planaren Brüchen (PF) auf, sind stark alteriert und manche sind teilweise aufgeschmolzen. Sekundärer Magnetit entstand nach dem Einschlag innerhalb des Suevits und bildet lockere Aggregate, die aus nadelförmigen Einzelkristallen bestehen, welche wiederum wenige nm bis ca.10 µm lang sind. Die kleinsten Körner in diesen Anhäufungen sind superparamagnetisch (SP) und weisen herabgesetzte Curie-Temperaturen (T_C) zwischen 520 und 560°C auf. Single Domänen (SD)- und MD Magnetitkörner sind durch eine T_C von 580°C gekennzeichnet. In vielen gesteinsmagnetischen Messungen des geschockten oder sekundären Magnetits ist die Temperatur des Verwey-Übergangs (T_V) erniedrigt oder unterdrückt. Dieses Verhalten ist auf eine nicht-stöchiometrische Zusammensetzung der Magnetite zurückzuführen. Eine solche Zusammensetzung entsteht überwiegend durch Oxidation, welche zuerst die Kornränder erfasst und dann größere Bereiche eines Korns einnimmt. Ob und wie stark T_V modifiziert wird, hängt dabei entscheidend von dem Verhältnis zwischen oxidertem und nicht oxidierten Volumenanteil eines Korns ab.

Neben Magnetit enthält der Suevit auch geschockten Pyrrhotin, der großteils ein starkes Eisendefizit aufweist. Geschockter Pyrrhotin hat eine Zusammensetzung, die ungefähr der von Smythit (Fe_9S_{11}) entspricht und wird in dieser Arbeit als Eisen-defizitärer Pyrrhotin bezeichnet. Dieser Pyrrhotin zeigt keinen oder einen stark unterdrückten 34 K-Übergang (monoklin zu trikliner Phasenübergang des 4C Pyrrhotins) und eine deutlich erhöhte T_C zwischen 340 und 365°C (T_C von 4C Pyrrhotin: 320°C). Außer häufig auftretenden Stapelfehlern weist dieses Mineral keine der Deformationsstrukturen auf, die in experimentell geschockten Pyrrhotin gefunden wurden. Da Stapelfehler auch ein typisches Merkmal von Smythit sind, wird geschlussfolgert, dass Alteration den geschockten Pyrrhotin gänzlich modifiziert hat. Das Eisendefizit und die Ausbildung einer geordneten magnetischen Struktur können somit auf Alteration und die damit verbundenen Diffusionsprozesse zurückgeführt werden. Da die Stapelfehler eine starke Unordnung im Kristall hervorrufen, entsteht der Ferrimagnetismus dabei höchstwahrscheinlich innerhalb von ca. 10 nm breiten, defektfreien Domänen.

Die Mikrostrukturen in geschocktem Pyrrhotin wurden während der Anfangszeit dieser Arbeit zunächst als ein Produkt der Stoßwellenmetamorphose interpretiert. Dem lag die Tatsache zugrunde, dass dieser Pyrrhotin bereits sehr stark alteriert ist. Es gestaltete sich daher schwer, die Mikrostrukturen, die mit dem Impaktereignis in Verbindung stehen, von denen zu trennen, die durch Alteration erzeugt wurden. Um dies besser zu verstehen, wurden daher Schockexperimente mit Drücken zwischen 3 und 30 GPa an Pyrrhotin durchgeführt. Diese Experimente zeigen, dass Mikrostrukturen, die durch das Schockereignis hervorgerufen, aber durch Nicht-Alteration verändert wurden, nicht mit denen des natürlich geschockten Pyrrhotins übereinstimmen. Pyrrhotin, welcher Drücke bis zu 8 GPa erfahren hat, ist überwiegend mechanisch deformiert. Diese Deformation beinhaltet eine Korngrößenverkleinerung, die Bildung von PFs, Planaren Deformationsstrukturen (PDFs) und Gitterdefekten. Damit verbunden ist eine Zunahme der Koerzivität und der remanenten Sättigungsmagnetisierung (SIRM), wobei diese Veränderungen generell ein erhöhtes Single-Dömänen-Verhalten (SD) hervorrufen. Schockdrücke zwischen 20 und 30 GPa produzieren dagegen amorphe Domänen im Kristallgitter und bei 30 GPa finden sich geschmolzene Partikel unter den Körnern. Die stoßwellenbedingte Aufschmelzung ist mit der Kristallisation von elementarem Eisen assoziiert.

Die Ergebnisse dieser Arbeit zeigen, dass Schockdrücke eine Reihe von strukturellen und magnetischen Veränderungen in magnetischen Mineralen hervorrufen. In einem natürlichen Umfeld werden viele dieser Merkmale jedoch durch Alterationsprozesse wieder verändert. Die gesteinsmagnetischen Eigenschaften von geologischen Einheiten in Impaktkratern sind daher von den Ausgangseigenschaften der magnetischen Träger sowie ihrer jeweiligen impaktbezogenen Deformations- und Alterationsgeschichte abhängig. In Impaktsedimenten ist zusätzlich die unregelmäßige räumliche Verteilung der magnetischen Minerale von Bedeutung, da eine lokale Anhäufung solcher Minerale in den jeweiligen Bereichen eine stärkere Magnetisierung erzeugt. Das magnetische Anomalienmuster einer Impaktstruktur ist daher nicht zwingend an die Kraterstruktur gebunden, sondern vielmehr das Ergebnis einer komplexen Interaktion unterschiedlicher klein- und großräumiger Prozesse.

Contents

Abstract		1
Zusammenfassung		3

1 General aspects — 7
- 1.1 Introduction — 8
  - 1.1.1 General aspects of impact cratering — 8
  - 1.1.2 The Chesapeake Bay impact structure — 10
  - 1.1.3 Objectives of this thesis — 12
  - 1.1.4 Structure of work — 13
- 1.2 Theory and methods — 15
  - 1.2.1 Theoretical background — 15
  - 1.2.2 Methods — 19

2 Results — 22
- 2.1 Rock magnetic properties of units from the CBIS — 23
  - 2.1.1 Introduction — 23
  - 2.1.2 Results — 25
  - 2.1.3 Discussion — 31
  - 2.1.4 Conclusions — 35
- 2.2 Microstructures and magnetic properties of pre- and post-shock magnetite in the CBIS — 38
  - 2.2.1 Introduction — 38
  - 2.2.2 Results — 40
  - 2.2.3 Discussion — 48
    - 2.2.3.1 Primary and shocked magnetite — 48
    - 2.2.3.2 Secondary magnetite — 51
  - 2.2.4 Conclusions — 56
- 2.3 Iron-deficient pyrrhotite in the suevite from the CBIS — 59
  - 2.3.1 Introduction — 59
  - 2.3.2 Results — 60
  - 2.3.3 Discussion — 68

		2.3.4	Conclusions	74
	2.4	\multicolumn{2}{l}{Shock experiments with pressures between 3 and 30 GPa on hexagonal and monoclinic pyrrhotite}	76
		2.4.1	Introduction	76
		2.4.2	Material and experimental setup	78
		2.4.3	Results	79
		2.4.4	Discussion	89
		2.4.5	Conclusions	96
3	\multicolumn{3}{l}{**Final conclusions**}	**98**
	3.1	\multicolumn{2}{l}{Iron-deficient and experimentally shocked pyrrhotite}	99
	3.2	\multicolumn{2}{l}{Comparison of shock-related features between magnetite and pyrrhotite}	102
	3.3	\multicolumn{2}{l}{Natural remanent magnetization}	103
	3.4	\multicolumn{2}{l}{General relevance for magnetic properties of impact structures}	104

References   107

Abbreviations   123

Chapter 1

General aspects

1.1 Introduction

1.1.1 General aspects of impact cratering

During the last 30 years it has been generally accepted that impact cratering is a fundamental geological process affecting nearly every planet and satellite in the solar system (Melosh and Ivanov, 1999). Whereas impact craters are well-preserved on many extraterrestrial planets like Moon or Mars (Ivanov, 2001; Stöffler et al., 2006), erosion and tectonic activities are successively blurring and deleting such structures on Earth. Furthermore, the registred impact record on Earth is mainly restricted to what is known from accessible land surface, but impact structures existing at Earth oceans floors remain widely undiscovered. Nowadays about 200 terrestrial impact structures are known (W.U. Reimold, personal communication, McCall, 2009). Impact structures are generally classified into different types depending on their geometry (e.g. French, 1998; Grieve and Pesonen, 1992; Reimold and Koeberl, 2008). Detailed and comprehensive studies on different types and formation mechanisms of impact structures have been published by French (1998), Melosh (1989), Melosh and Ivanov (1999) and Reimold and Koeberl (2008).

The main impact structure types comprise simple, complex, and multi-ring craters. Simple craters have a bowl-shaped geometry and show an average depth-diameter ratio of ~1:3, whereas complex craters are characterized by the presence of a central peak and modified crater rims, which consist of slumped blocks forming so-called terraces. Multi-ring craters contain two or more ring-shaped elevations forming a collapsed central peak. The crater morphology is dependent on the impactor size, but also on composition, porosity and water-content of the target material (Koeberl, 1994; Wünnemann et al., 2006). Strong gravity and high porosity generally reduce the crater size, whereas water in the rocks creates the strong opposite effect. Terrestrial simple craters are observed in sedimentary or crystalline rocks with approximate crater diameters of 2 or 4 km, respectively (Reimold and Koeberl, 2008). Complex, or peak-ring structures range from ~2 to more than 100 km in diameter. Examples for the different crater types on Earth are the Barringer crater, Arizona, USA (simple crater), the Chesapeake Bay impact structure (CBIS), Vrigina, USA (complex crater) and the Chicxulub impact structure, Mexico (multi-ring crater). The results from this study are based in findings from the CBIS.

The general understanding of impact crater formation processes is based on the study of shock wave physics and the knowledge of material properties of the particular types of rock involved in the cratering process. A good summary of shock wave physics has been composed by Langenhorst (2002) Further detailed descriptions on this topic can be found in Duvall and Fowless (1963) and Melosh (1989).

In general, a shock wave is produced by longitudinal (compressional) waves with high amplitudes and velocities. In contrast to elastic waves, the shock wave front is a sharp borderline behind which compression is extremely high. Compression induces a certain shear stress in the affected rocks and when the Hugoniot elastic limit of a rock is reached, material failure results in brittle or ductile deformation. Starting at the shock centre, the shock wave propagates in all directions. When it reaches the free surface, a release (rarefaction) wave is triggered. The release wave runs in the opposite direction and catches up with the precursory shock front after some time as the wave propagation is accelerated in the compressed and denser material. When the release wave and the shock front start to temporally overlap, an "equilibrium" state is established, consisting of a compressional shock front which is followed by subsequent pressure release. After the shock wave has passed through the material, a certain particle velocity remains in the material. This "rest-energy" is a result of the particle movements which take place during compression and decompression. It induces vaporization, melting, fracturing and ejection of material (excavation).

Impact cratering is different to usual geological processes since the basic crater formation occurs within a few minutes (Melosh, 1989). Concerning the development of an impact crater, three major stages have been defined. An outline of the principal impact crater formation stages for the CBIS, is illustrated in Fig. 1.1.1. First, the "contact and compression stage" defines the earliest crater stage, beginning at the contact between projectile and target material, and ending when compression has entirely affected the

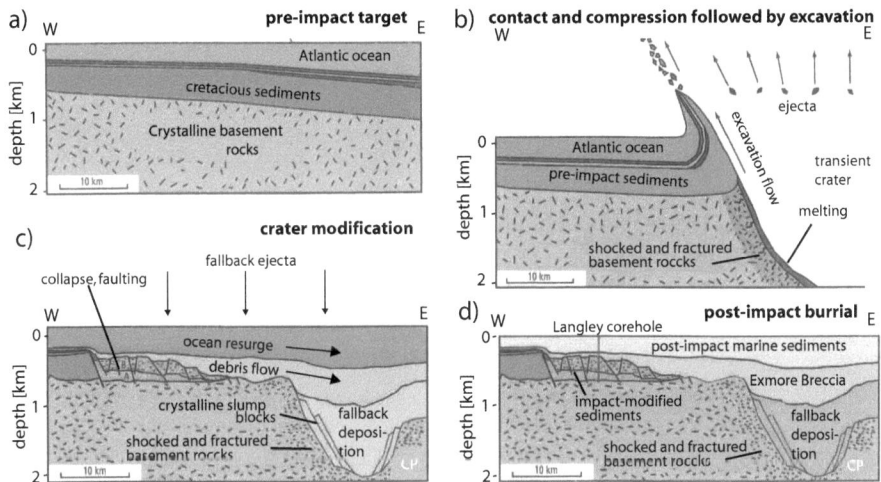

Figure 1.1.1: Schematic crater formation history of the Chesapeake Bay impact structure (CBIS). a: Setting of the pre-impact target. Three main layers defined the continental shelf setting at Chesapeake Bay previous to the impact: crystalline basement rocks, siliciclastic cretaceous sediments and ocean water. b: Contact and compression (not shown) caused melting at the contact zones and shock metamorphism in the target. Excavation generally begins when the kinetic energy is completely transferred to the target. A transient crater forms by expansion of the shock wave. c: Crater modification occurs when the maximum transient crater size is reached and is a consequence of gravitational collapse. When the central peak (CP) developed in the CBIS, a violent ocean water resurge filled the crater with water and fragmented rock material. Sediments at the outer parts of the crater collapsed and formed tilted blocks. d: post-impact burial was finished after deposition of the entire crater fill and the marine post-impact sediments. Modified from Edwards and Powars (2003).

impactor (Reimold and Koeberl, 2008). The extremely high strain rates produce vaporization, melting and shock metamorphism. Early crater products are found as tektites in large distances to the crater, or within the crater as impact melt or suevites. Second, the "excavation stage" begins once the kinetic energy of the projectile has been entirely transferred to the target. Particle movement, induced by the passage of the shock wave, is first directed downwards and inwards, but subsequently turns upwards as a consequence of the decompression process (Stöffler et al., 2006). The cratering process is stopped by the lithostatic pressure, which is acting against crater growth. Crater rims are defined by zones of ductile and brittle deformation, where stress exceeded the strength of the material. The main part of the ejecta is deposited between this stage and the final "crater modification stage". The latter is defined to begin when the maximum depth of transient crater is reached (Reimold and Koeberl, 2008). Here, gravitational collapse forms the final crater morphology and in complex craters, the central peak develops. In the case of Chesapeake Bay, the crater modification was accompanied by a large ocean water resurge process which filled the open cavity with water and debris (Horton et al., 2005a). Impact metamorphic rocks are generally described as "impactites" and occur as crater fill. Ejecta, on the other hand, are deposited in distance to the crater. The crater fill mainly comprises highly unsorted, polymict breccia, impact melt, pseudotachylites and suevites. Suevite usually forms a polymict impact breccia containing clastic and melt particles (Reimold et al., 2012; Stöffler, 1971; Stöffler and Grieve, 1994).

1.1.2 The Chesapeake Bay impact structure

The late Eocene (35.3+/- Ma; Horton and Izett, 2005) Chesapeake Bay impact structure (CBIS) has been formed by the collision of a ~3 km diameter meteorite (Collins and Wünnemann, 2005) with the continental shelf of the Atlantic margin (Fig. 1.1.2). The impact formed a complex crater of ~85 km diameter (Horton et al., 2005c) and is supposed to be the source of the North American tektite strewn field (Deutsch and Koeberl, 2006; Koeberl et al., 1996). The recent structure is one of the best-preserved impact structures on Earth, but the deposition of ~500 m of coastal plain sediments (Horton et al., 2005c) onto the structure makes it accessible for scientific studies only by geophysical survey and drillings (Gohn et al., 2009b). The structural features have been mainly deduced from seismic surveys (Catchings et al., 2008), gravity- (Plescia et al., 2009) and magnetic anomalies (Shah et al 2005, 2009), and were complemented by several drillings into the crater. According to the current research, the significant target rock strength variations (weak sedimentary units underlain by much stronger crystalline basement) and the wet conditions under which the excavation took place, explain the main features of the ultimate shape, size and the internal stratigraphy (Gohn et al., 2006, 2008). The crater fill consists of crystalline basement, which is overlain by impact sediments. The latter are intercalated with melt, crystalline fragments and megablocks (Horton et al., 2009a).

This thesis is based on the sample material of four different boreholes, which were drilled into the CBIS. Their locations are marked in Fig. 1.1.2. Magnetic measurements from one of them, the Eyreville core, have been previously published by Elbra et al. (2009) and Shah et al. (2009). These measurements are complemented by various measurements performed in this study. Samples from the other three drill cores were obtained by the author during a sampling campaign in July 2009. The samples were chosen on the basis of the results given in Elbra et al. (2009) who found that only suevite and crystalline basement contribute significantly to the bulk magnetic anomaly field over the CBIS. In the following, a brief summary on each drilling and the associated drill core profiles will be given. Detailed descriptions of the geologic, petrographic, geochemical and geophysical analyses can be found in Horton et al. (2005c) and Gohn et al. (2004, 2009a). For lithological profiles see Fig. 2.1.2 - 2.1.4.

The Eyreville drilling is the deepest borehole which was drilled into the crater and has been accomplished between September and December 2005 by the International Continental Scientific Drilling Program (ICDP) and the U.S. Geological Survey (USGS). The continuously cored borehole reached a final depth of 1766 m and was drilled above the circular structural low, in short distance to the crater's central uplift. This area shows a high gradient between a magnetic low and high. From seismic images it seems likely that the drilling did not reach the autochtonous basement (Catchings et al., 2008, Powars et al., 2009). The core profile is composed as follows: Below a 444 m thick layer of post-impact sediments of Eocene to Pleistocene age, a large section of a polymict sediment-clast-dominated impact breccia (Exmore breccia) occurs with a thickness of 652 m. This section contains only mi-

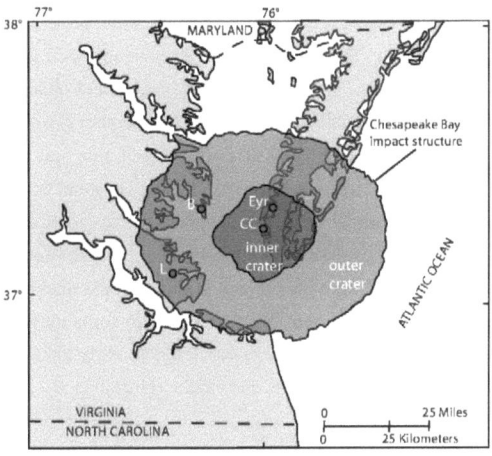

Figure 1.1.2: Location map of the Chesapeake Bay impact structure (CBIS) in Virginia, USA. The inner and outer crater is highlighted in dark and light blue, respectively. Red dots mark the four drill sites from which samples for this thesis have been studied. Eyr: Eyreville drill site; CC: Cape Charles drill site; B: Bayside drill site; L: Langley drill site. Modified from Gohn et al. (2006) and Powars et al. (2009).

nor amounts of impact melt. The breccia is followed by a granite megablock of 275 m thickness that has presumably been detached from the neighbouring basement and transported into the crater centre during crater wall-collapse (Kenkmann et al., 2009). The granite block shows no evidence of shock metamorphism and consists of two main granite types: a mostly fine-grained biotite granite of Neoproterozoic age and an underlying medium-coarse-grained biotite granite of Permian age. Below, a thin interval (25 m) occurs. This section contains quartz sand and lithic rocks. Underneath a section of suevite is following. It consists of impact melt, lithic breccia and basement derived blocks. Multiple sets of planar deformation features (PDF), evidencing shock, (Grieve et al., 1990; Goltrant, 1992; Langenhorst, 1994) have been found in quartz occurring in the matrix or intercalated rock fragments (Horton et al., 2008). Finally, a 215 m thick graphite-sillimanite-mica-schist megablock (Catchings et al., 2008) appears at the bottom of the drill core. This rock has experienced metamorphism in the upper amphibolite facies. It is partially mylonitic and is intruded by granitic and pegmatitic material (Townsend et al., 2009), which dominates the section in the lower part. Alteration can be recognized by a secondary mineral assemblage comprising sericite, epidote and calcite (Townsend et al., 2009).

The Cape Charles borehole was drilled by the USGS during May and June 2004 to a total depth of 823 m as a pilot hole and precursor of the Eyreville drilling. The drill site is located above the central peak (see Fig. 1.1.2). However, this partly cored drilling did not reach the basement rocks (Catchings et al., 2008). Three main units were distinguished from the profile section (Gohn et al., 2009b, Horton et al., 2009a): (1) crystalline-clast breccia and quartzo-feldspathic cataclastic gneiss that form a lower impact unit between 655.3 and 822.6 m depth, (2) a sediment-clast breccia that occurs between 354.5 and 655.3 m depth and forms an upper impact unit, and (3) post-impact sediments that appear between land surface and a depth of 354.5 m. The breccia at the bottom of the lowest section partly contains significant amounts of impact melt and is therefore classified as suevite. Gneiss is associated with small amounts of chlorite-rich mafic rocks (Gohn et al., 2004; Horton et al., 2005a). Samples used in this study were obtained both from the upper unit that contains suevite, and from basement blocks. The latter consists of few gneiss blocks and megablocks (up to some tens of m) as well as of an amphibolite and mafic dyke block. Melt and mafic dyke block are strongly altered and contain secondary minerals such as chlorite, white mica, quartz, albite and calcite (Horton et al., 2006).

The Langley drilling was accomplished as a cooperation project between NASA and the USGS between the years 2000 and 2002 down to a final depth of 635.1 m (Horton et al., 2005b). Post-impact sediments and crater fill units make up the main part of the cored material. Crystalline basement was encountered below 626.3 m and consists of an Neoproterozoic monzogranite. This peraluminous rock is pervasively chloritized and nonfoliated.

The Bayside drilling was conducted to a total depth of 728.5 m (Horton et al., 2002). Below the crater fill, which ends at a depth of 708.9 m, unshocked granitic basement rocks were drilled. These rocks are pale red and consist of medium-grained, non-foliated and highly chloritized monzogranites. An unmapped subcrop suture is proposed by Horton et al. (2002) to separate the terrane sampled in both the Langley and Bayside core from Mesoproterozoic Laurentian basement beneath the Coastal Plain of New Jersey. Bayside and Langley granites lack of any evidence of shock or discernible heating.

1.1.3 Objectives of this thesis

This thesis is part of the DFG research project "Rock magnetic properties and their anisotropy from host rock and impact lithologies of drillings at the Chesapeake Bay impact structure, USA". First rock magnetic measurements on different lithologies of the Eyreville drill core were done in 2008 in order to characterize their rock magnetic properties. These results were published together with data from Dr. Tiiu Elbra from the University of Helsinki, who conducted complementary measurements on rocks from this crater structure, in Elbra et al. (2009). Based on these results, a second sampling campaign was accomplished in 2009 at the USGS in Reston, USA, for core material from the Cape Charles, Bayside and Langeley drillings. Material from all four mentioned drill cores was then subjected to rock magnetic measurements (chapter II.3). The main focus of this thesis is a detailed magneto-mineralogical study. It aims to understand how impact-related processes modify rock magnetic properties if magnetite and pyrrhotite, the two most important magnetic minerals in the Earth crust, are involved.

During this study, a series of striking and up to now unknown magnetic features was observed in the suevite of the CBIS. Different generations of magnetite (pre- and post-shock) and pre-shock pyrrhotite have been investigated in detail. Additionally to various kinds of magnetic measurements and experiments, this study involved the determination of characteristic microstructures with the help of optical microscopy, scanning electron microscopy, transmissions electron microscopy and electron microprobe analysis. A crucial finding was the strong iron-deficiency in pyrrhotite and associated magnetic features, which have been described in detail for the first time in association with an impact structure. Strong alteration of the grains, however, hampered the interpretation and it remained unclear if iron-deficiency was a pre-shock feature induced by shock or subsequent alteration. Therefore, natural pyrrhotite ore was shocked between 3 and 30 GPa using an air gun and high explosive devices. The well-defined conditions of the experiments allowed the examination of pure shock-related features of pyrrhotite. The main scope of this study was therefore the examination of shock-induced microstructures in pyrrhotite and magnetite and their consequences on the particular magnetic behaviour.

1.1.4 Structure of work

This thesis consists of three main chapters. The first comprises a general introduction and a brief explanation of the theory and methods. The second chapter focuses on the results of this study. It contains four independent sections, each of which consisting of an abstract, introduction, results, discussion and conclusion part. The last chapter gives some final conclusions and discusses the results of this thesis in a more general context. All raw data produced during this study are given in the appendix on a CD.

Chapter 2.1 presents and discusses the entire data on the rock magnetic properties of units from the CBIS. It gives insight into the remanence types of the particular units and their stability.

Chapter 2.2 describes and discusses the different types of magnetite occurring in the studied rock units. General structural and magnetic features are discussed and a formation history of each type is developed.

Chapter 2.3 shows the structural and magnetic peculiarities of naturally shocked pyrrhotite from the CBIS. The results are compared with general features described for shocked pyrrhotite in the literature.

Chapter 2.4 describes the results of the shock experiments and gives a general overview of microstructures and magnetic features, which are characteristic for shocked pyrrhotite.

1.2 Theory and methods

This section gives an brief summary of the basic knowledge necessary to understand the main results and interpretations of this study. While the first part consists of an inroduction to the main rock magnetic parameters, the second part lists the applied measurement instruments. Moreover, the particular measurements will be shortly outlined.

1.2.1 Theoretical background

Fundamentals of rock magnetism

At the latest since the invention of the compass, magnetism has become a fascinating and important phenomenon for mankind. The word "magnet" comes from the Latin and Greek word "magnes", derived from a piece of lodestone, which is a naturally occurring ore of magnetite (Fe_3O_4) first described in several fables in 1000 B.C. After analyzing the field lines around a spherical lodestone and the field lines of the Earth, William Gilbert, a physician to Queen Elizabeth I of England, concluded around 1600 AD that the Earth itself is a great magnet. Later on, Johan Gauss found that the bulk magnetic field was generated within the Earth and is forming a dipole field. The fundamental groundwork of Johan Königsberger, Takesi Nagata and Emile Thellier, who discovered the acquisition process of a thermoremanent magnetization (TRM), finally allowed the correct interpretation of the magnetic stripe pattern on the ocean sea floor, which was discovered in the 1950s. This led to the development of the model of sea-floor-spreading which turned out to be crucial for the acceptance of the up to then hardly debated theory of plate tectonics. Today, magnetic anomalies are widely used in order to map geologic structures or metamorphic terranes, although the connection between the magnetic field and the linked parameters is not always obvious. This is in particular true for the magnetic anomaly pattern over the CBIS which results from the complex crater built-up and numerous pre-, syn,- and post-impact modifications of various magnetic mineral assemblages. However, to understand the rock magnetic study of this thesis, an introduction to the most important fundamentals of magnetism is indispensable and will be given in the following. Detailed textbooks about rock magnetism have been written by Soffel (1991), Dunlop and Özdemir (1997) and Butler (1998).

The magnetization of a material, when an external magnetic field is applied, depends mainly on three parameters which are linked by the equation
$$H = \chi \cdot J,$$
where H represents the external magnetic field, χ the magnetic susceptibility (for rock magnetic purposes usually related to a specific volume) and J the magnetization. χ is a material constant and is thus a parameter for the magnetizability of a substance which determines the amount of induced magnetization, J_i. The magnetization remaining in the material after removal of an external magnetic field is the remanent magnetization (M_{rs}). The Königsberger ratio Q allows one to estimate the dominance of the

induced or remanent magnetization through the relation:

$$Q = \frac{M_{rs}}{\chi \cdot H}$$

When an electric current is flowing through an inductor a magnetic field is induced parallel to the inductor axis. In the same way an electron produces a magnetic field during its motion around the atomic nucleus. Each electron holds an effective magnetic spin moment, resulting from its intrinsic angular moment (spin), and the orbital motion. The direction of the magnetic moment is indicated by a magnetic field vector. If all electrons in an atom have paired spins, there are as many spins oriented in one direction as there as spins oriented in the opposite direction. In that case the overall magnetic moment of the medium vanishes. Materials without net magnetic moments are called *diamagnetic*. Atoms containing one or more unpaired electrons have a certain net magnetic moment (average spin magnetic momentum of an electron (Bohr magneton): $\mu B = 9.27 \cdot 10^{-24} Am$) and are referred to as *paramagnetic*. If an external magnetic field is applied, the magnetic moments are aligned in direction of the magnetic field. However, removal of the magnetic field results in randomization of the particular orientations. This is the reason why paramagnetic materials have no permanent magnetic moment. However, in some materials strong interactions of paramagnetic moments result in a remaining magnetic ordering. Such materials are called *ferromagnetic* and have a permanent magnetic dipole moment. In ferromagnets, spin ordering is happening without application of an external field. One important property of ferromagnets is that this stable arrangement is temperature-dependent. Above a critical temperature, the Curie temperature (T_C), the spins disorder and the material becomes paramagnetic. In the stable state, several arrangements of the magnetic moments are possible within a crystal lattice. In an *antiferromagnetic* ordering, the magnetic moments of the individual atoms are aligned in an antiparallel way resulting in a zero net magnetic moment of the whole crystal. Such an arrangement is typical for oxides, where different spin-spin interactions occur between the oxygen atoms. Small net magnetic moments can result if the spins in an antiferromagnetic ordering are not precisely antiparallel, but slightly canted, which is why the term *canted antiferromagnestism* has been introduced. The strongest net magnetic moment results from a *ferrimagnetic* arrangement, within which the spins are antiferromagnetically ordered, but unequal strengths of those spins produce a net magnetic moment in one direction. In all those magnetic arrangements, parallel coupling of spins from neighbouring atoms occurs within the so-called *magnetic domains*. The latter describe regions inside the material where all individual spin momenta are pointing in the same direction. Ferrimagnetic materials have net magnetic moments due to antiparallel coupling of unequal spin momenta inside these domains.

In general, three different magnetic domain states are distinguished in rock magnetism. The occurrence of a particular domain state is mainly dependent of the grain size and shape. The term *multidomain* (MD) grains denotes grains which contain multiple magnetic domains. These domains are uniformly magnetized, but differ in their bulk vector orientation from domain to domain. If a magnetic field is applied, a net magnetic moment is produced which is mainly due to the movement of domain walls.

Magnetic domains with vectors aligned approximately in direction of the applied field grow in size at the expense of other domains. However, after removal of the field domains move back, close to their initial positions. Because domain walls are particularly stable on certain energetically favourable positions (due to lattice imperfections and internal strains), a small net magnetic moment remains. However, even weak magnetic fields are sufficient to remove the walls, which is why MD grains are in general magnetically soft. *Single domain* (SD) grains are grains which are small enough so that only one uniform magnetic domain is present. Such grains are very efficient magnetic remanence carriers. The high remanence stability mainly arises from the fact that the only way to change magnetization in such grains is by rotating the remanent magnetization M_{rs}. This process is much harder to achieve than the movement of magnetic domain walls. *Pseudo-single-domain* (PSD) grains are grains with sizes between the ones of MD and SD grains. Such grains usually have a small number of magnetic domains. In general, grain associations showing a PSD-like behaviour are often assemblages containing SD grains and MD grains with few magnetic domains. Finally, *superparamagnetic* (SP) grains are single domain grains which are too small to preserve a magnetic moment since their characteristic magnetic *relaxation time* (τs) is <100 s. Magnetic relaxation is referred to as the natural exponential decay of remanent magnetization in SD grains with time.

The presence of magnetic domains reduces the overall free energy associated with the magnetic ordering. Within a particular magnetic domain a uniform magnetization is present, but this magnetization varies between the individual domains. In order to describe the behaviour of magnetic ordering, three important energy terms are introduced. First, magnetocrystalline anisotropy determines the direction of magnetization, since one particular crystallographic direction is easier to magnetize than others. This direction is called *easy direction of magnetization* or *easy magnetic axis*. Second, magnetostatic effects arise from Coulomb interactions between the domain boundaries or free surfaces. The resulting magnetostatic energy depends on the crystal shape and magnetic domain distribution and is minimized when the magnetic moments in the particular domains are aligned in opposite directions. Third, *magnetostriction* describes the phenomenon that the shape of a magnetic mineral changes during magnetization. This effect is mainly ascribed to the shift and rotation of magnetic domains during the magnetization process. All three terms influence the magnetic domain structure, which is generally arranged such that the overall free energy is at minimum.

Magnetic domain states are usually characterized by four important parameters. During so-called hysteresis measurements, a magnetic field is applied and successively increased until all moments are aligned along the field direction. Then, the *saturation remanence* (J_s or M_s) is reached. After reducing the field to zero again, a *remanent magnetization* (M_{rs}) remains in the mineral. In order to completely destroy this magnetization ($M_{rs} = 0$) an opposite field has to be applied. The corresponding field strength is referred to as the *coercive force* (H_c), whereas the field necessary to reorient one half of M_{rs} into the opposite direction is called *remanence of coercivity* (H_{cr}). Both parameters are expressions of the *coercivity*, which is a measure of the particular magnetic hardness.

Natural remanent magnetization (NRM) is the remanent magnetization in a rock and has usually several components depending on the rock history. The particular magnetizations are called *characteristic magnetizations* (ChRM) and usually consist of one of the following magnetization types: *Thermoremanent magnetization* (TRM) is acquired when a magnetic mineral is cooled down to a temperature below its particular T_C, a process usually occurring in magmatic rocks. A TRM representing the direction of the present Earth magnetic field is then preserved in the crystal. Chemical modifications of the minerals present and precipitation of new minerals from solutions produce a *chemical remanent magnetization* (CRM). The basic underlying process is the growth of secondary minerals which acquire a stable magnetization when their grains size exceeds a certain threshold size. If rocks are exposed to shock, they can loose large parts of their present magnetization. A new *shock remanent magnetization* (SRM), representing the magnetic field present at the time of shock, can then be acquired afterwards. Thermal fluctuations can be responsible that a mineral acquires a *viscous remanent magnetization* (VRM), which partially modifies the original NRM. Finally, drilling can impose a *drilling induced magnetization* (DIRM) on rocks representing a magnetic field which is produced from the magnetized core barrel.

Magnetic minerals

MAGNETITE (Fe_3O_4) is the end member of a solid solution with ulvospinell (Fe_2TiO_4) called titanomagnetite. It has an inverse spinell structure with oxygen atoms forming a face-centred cubic lattice and Fe^{2+}- and Fe^{3+}-cations occupying the interstitial sites. Whereas Fe^{2+}-ions are restricted to octahedral sites, Fe^{3+}-ions occupy both tetrahedral and octahedral sites. Different magnetic moments arising from both sublattices (Fe^{3+}: 5 µB; Fe^{2+}: 4 µB) leave one net magnetic moment which is resulting from antiparallel coupling. At room temperature (RT) the easy magnetic axis is along [111]. Yet, when cooling down, at ~130 K an isotropic point (T_i) is reached at which magnetite is isotropic. Below the Verwey transition (T_V) at ~120 K, the crystal structure becomes monoclinic. T_C of pure magnetite appears at 580°C.

PYRRHOTITE ($Fe_{1-x}S$) is a solid solution between FeS (troilite) and Fe_7S_8 (4C pyrrhotite). The structure of all pyrrhotite modifications is based on a NiAs subcell and can be described as cation layers which alternate with hexagonally close-packed S-layers. The whole sequence is stacked along the c-axis. Compositional deviations from the FeS structure require the introduction of vacancies occupying some of the octahedral sites and resulting in a slight distortion of the hexagonal unit cell. The vacancies are arranged in such a way that maximum separation can be achieved. Therefore, only every second cation layer contains vacancies. Within a cation layer, magnetic moments are ferromagnetically coupled, but negative exchange coupling in the adjacent layers neutralizes the magnetic moments. However, the vacancy arrangement in the 4C structure distorts the structure into monoclinic and produces a net magnetic moment. At room temperature this arrangement produces the only ferrimagnetic modification within the pyrrhotite group. It becomes paramagnetic above its T_C of 320°C. At 200°C, antiferromagnetic hexagonal (NC) pyrrhotite transforms into ferrimagnetic (NA) pyrrhotite. This transition is called λ-transition and is based on thermally activated vacancy ordering. Similar to the T_V in magnetite, monoclinic pyrrhotite shows a low-temperature magnetic transition at 30 - 34 K which is proposed to be a crystallographic transition at which the structure changes from monoclinic to triclinic symmetry.

1.2.2 Methods

A broad range of measurements have been applied in the framework of this study and some were done in cooperation with other working groups. Electron microscope and microprobe measurements were conducted with the help of the particular staff members from the mentioned research institutes. Measurements at Bremen University were entirely done by the local workers.

Magnetic measurements

Magnetic susceptibility measurements as a function of temperature (χ-T) are generally useful to relatively quickly determine the magnetic carriers; mainly by means of their particular magnetic transitions. Such measurements have been conducted on a KLY-4S kappabridge (300 Am^{-1} and 875 Hz) combined with a CS-3 and CS-L furnace apparatus of AGICO. Cooling was achieved with the help of liquid nitrogen and cooling/heating rates for this tool range between 3 and 4°C/min for the low-temperature, and between 11 and 14°C/min for the high-temperature runs. The latter were done in an argon atmosphere to avoid oxidizing mineral reactions during heating (flow rate of 110 mL min^{-1}). The raw data were corrected for the empty cryostat/furnace and normalized with respect to the sample weight. T_C was determined by the inverse susceptibility method described in Petrovský and Kapička (2006). Measurements of natural remanent magnetization (NRM) were done with a JR5A spinner magnetometer (AGICO Company). Alternating field (AF) demagnetization was performed in peak fields up to 160 mT with a MI AFD 1.1 from Magnon International. Stepwise thermal demagnetization in increments of 50°C up to 700°C was done with the Thermal Demagnetizer MMTD1 (Magnetic Measurements). The remanence and susceptibility experiments are generally preformed using crushed rock powder. During AF and thermal experiments, a certain field, respectively temperature is applied on the sample and after removal of the particular parameter, remaence is measured. This procedure is repeated while the magnetic field, respectively temperature is successively increased. The demagnetization experiments require samples with an undistorted sample fabric in order to preserve the magnetic directions therein. Therefore, cylinders with a defined volume were drilled from the provided drill core halfs. These cylinders have a diameter of 1.4 cm and a height of 1.2 cm resulting in a total volume of 1.8 cm^3. Since usually a standard cylinder volume of 10 m^3 is used for rock magnetic purposes, the values of all measured parameters have been converted to this volume after measurement. All yet described procedures were conducted at the Institut für Angewandte Geowissenschaften, Karlsruher Institut für Technologie (KIT), Karlsruhe, Germany. Low temperature (LT) magnetic measurements are usually conducted between 10 and 300 K (room temperature: RT) and either remanence (DC) or susceptibility (AC) is measured as a function of temperature. Such measurements are primarily used for magnetic mineral identification based on low-temperature crystallographic transitions and for characterizing particle size distributions. In this study the effect of shock on the particular transitions was an additional important point. The LT measurements were conducted at the Institute for Rock Magnetism (IRM), Minneapolis, USA with a MPMS 2 SQUID magnetometer system. The 20 and 30 GPa samples of the shock experimental setup described in chapter 6, underwent the same measurements on a Quantum Design MPMS XL7 SQUID magnetometer at the Institute of Marine Geophysics, Bremen University, Germany. Various LT remanence measurements were conducted on a set of selected samples. During RTSIRM measurements, saturation isothermal remanent magnetization (SIRM) was given at room temperature, cooled in an applied field down to 10 K and warmed back to room temperature. During zero field cooled measurement (ZFC), the sample was cooled in a zero field to 10 K and then a magnetic field was applied. Remanence was subsequently measured during warming after the field had been switched off. Finally, the sample was cooled in an applied field (FC), which was then switched off at 10 K and the sample was warmed back to room temperature when remanence was measured accompanying. The applied field for all experiments was 2.5 T. Recording of hysteresis loops allows determination of crucial mineral specific parameters. The most important are crucial parameters coercivity (H_c), remanence of coercivity (H_{cr}), saturation magnetization (M_s) and remanence of saturation magnetization (M_{rs}). During hysteresis, a magnetic field starting at 0 T is successively increased while magnetization is measured until the maximum field at which the sample is usually magnetically saturated (in measurements of this study: 1.7 T). Subsequently, the field is continuously decreased until the reverse maximum field is reached. Then, the field is again flipped and the procedure repeated. These measurements are also possible at low temperatures. The sample is then successively warmed from 15 K to RT while hysteresis loops are continuously recorded. The RT and LT hysteresis measurements were conducted on a Princeton Measurements vibrating sample magnetometer (VSM) at the IRM. This tool has a sensitivity of $5 \cdot 10^{-9} Am^2$. The 20 and 30 GPa samples decribed in chapter 6, were recorded on the Quantum Design MPMS XL7 SQUID magnetometer at University of Bremen. First order reversal curves (FORC) were equally recorded on the VSM at the IRM. FORC diagrams are calculated by addition of various "unfinished" hysteresis loops. The sample is first saturated and then a reversal field (H_a), which is less than the reverse saturation magnetization, is applied. The FORC is then the curve resulting when the field is applied from H_a back to M_s. This procedure is repeated for several values of H_a and from the resulting FORCs, various diagrams can be calculated. These measurements are helpful for determination of grain size distributions of magnetic minerals in a sample.

Microstructural measurements

Polished thin sections were prepared for various samples and analysed under reflected and transmitted light microscopy. In order to visualize the magnetic carriers, the thin sections were coated with a magnetic fluid, which contains ultrafine particles of magnetite. The ferrofluid (produced by Institut für Angewandte Polymerchemie, FH Aachen) sticks at grain or domain boundaries that generate a magnetic field. These parts then obtain a brown-reddish color. Scanning electron microscopic studies are based on an electron beam, which is moved over the sample surface and induces interactions with the sample material. The most common measurements analyze the secondary (SE) and backscattered (BSE) electrons, which give information about topography and material contrast, respectively. SEM images of selected samples were recorded on a LEO 1530 Gemini instrument. This measurement was accompanied by simultaneous energy-dispersive X-ray (EDX) analyses, which were performed with a Noran System 6 Thermo Fischer. EDX analyses allow determination of chemical compositions. Few samples were prepared for transmission electron microscopic (TEM). During TEM measurements, an electron beam is transmitted through an ultrafine sample lamellae. Such a lamella is produced with the help of a focused iron beam (FIB) instrument. Diffraction of the accelerated electrons is dependent on the atomic number of the elements in the sample and the so-called "diffraction pattern" is recorded at the end of the beam projection. Diverse setting possibilities of various apertures allow obtaining information on the crystallographic nature, but also to display microstructural features as dislocations. FIB preparation was conducted using an FEI Dual Beam Strata 400S at 30kV with a Ga^+ cathode. TEM studies on experimentally shocked pyrrhotite were conducted on an FEI Titan^3 80 - 300 microscope at 300kV accelerating voltage. Additional measurements on natural samples were achieved on a Philips CM 200 FEG at 200kV. The described measurements were all conducted at the KIT. Additionally, two samples from the Eyreville core (CB 29 and CB 32, discussed in chapter 2.3) were prepared and analyzed by TEM at the BGI in Bayreuth. For FIB preparation, a FEI Quanta 3D FEG (30 kV Ga^+) instrument was used and the specimens were examined using a 200 kV Philips CM 20 FEG TEM. Microprobe analyses allow determining the specific elements with the help of specific x-rays that are emitted when the sample is irradiated with an electron beam. Standards are required to allow quantitative conclusions of the particular composition. Such analyses were accomplished at the Museum für Naturkunde, Berlin on a JEOL JXA8500F field-emission cathode instrument at 15 kV. Marcasite (FeS_2) and pentlandite (($FeNi)_9S_8$) samples from Astimex Scientific Limited were used as standards.

Chapter 2

Results

2.1 Rock magnetic properties of units from the CBIS

Abstract

The CBIS is characterized by an irregular magnetic anomaly in which the regional NW-SE trending magnetic field pattern is disturbed by various steep short-wavelength (<2 km) anomalies. The anomalies do not define the crater rim but are interpreted to indicate shallow magnetic sources within the crater fill. These sources are most likely basement-derived megablocks which are embedded in the impact sediments. With respect to their magnetic properties, the granite, schist and gneiss megablocks are the most important ones. Together with the suevite, granite and schist are the only rock units which carry remanences above 0.1 A/m, but only schist and suevite carry a stable remanence. However, magnetic properties of the suevite are not homogeneous, since the particular magnetic carriers vary locally. Whereas secondary magnetite and shocked pyrrhotite occur in the Eyreville suevite, shocked magnetite is exclusively present in the Cape Charles suevite. The latter is characterized by an unstable magnetic remanence direction and rather low susceptibility values. Unstable remanence directions are also present in the Eyreville suevite, but in addition, a broad range of inclinations occur in those samples with stable remanence directions. The results indicate that stable directions are either thermo- or chemical remanent magnetizations. Parts of the results presented in this chapter have already been published by Elbra et al. (2009), but were completed by this thesis.

2.1.1 Introduction

Since the upturn of impact research in the early 70s, the study of terrestrial impact cratering processes has provided a large data basis that helps to develop models for cratering mechanics on Earth, but also on extraterrestrial bodies as Moon or Mars (Kenkmann and Schönian, 2006; Shoemaker, 1977; Stöffler et al., 2006). Terrestrial impact craters, however, have severely been modified by weathering (e.g. Melosh, 1989), if not subsequently been covered by post-impact sediments. In the latter case the original crater morphology is usually well-preserved, but information about its structural features and the particular impact lithologies is only available by drill holes and geophysical surveys. The three common geophysical methods comprise seismic, gravity- and magnetic data acquisition. Such measurements provide details about the crater excavation and collapse, and the deposition of impact sediments (Reimold and Koeberl, 2008). The gravity and magnetic anomalies are generally explained by brecciation and the formation of melt (Pilkington and Grieve, 1992). While the gravity anomaly reasonably reflects the crater morphology in most cases, the magnetic anomaly patterns are more variable and complex (e.g. Henkel, 1992; Plado, 2000; Urrutia-Fucugauchi, 2004). The most prominent feature over impact craters is a magnetic low (Clark, 1983; Therriault et al., 2002) which can be best recognized in crystalline environments (Pilkington and Grieve, 1992). Short-wavelength anomalies often modify such magnetic lows in complex craters (Shah et al., 2005) and some large structures (>40 km) are reported to exhibit central high-amplitude anomalies with dimensions of less than half of the crater diameter (Pilkington and Grieve, 1992; Pohl and Angenheister, 1969). However, many investigations of terrestrial impact structures demonstrated that a natural remanent magnetization (NRM) is the main cause of the observed magnetic anomalies and that the NRM of shocked rocks depends on the distance to the explosion (e.g. Gattacceca et al., 2007; Ugalde et al., 2007). The magnetic signature of impact craters is generally described to result from: a) compositional and magnetic properties of the target rocks, b) modification of magnetic minerals due to high, impact-related p-T conditions, and c) the natural remanent magnetization (thermoremanent magnetization (TRM), chemical remanent magnetization (CRM) or shock remanent magnetization (SRM)).

The Chesapeake Bay impact crater (CBIS) is characterized by an irregular magnetic anomaly in which the regional NW-SE trending magnetic field pattern is disturbed by various steep short-wavelength (<2 km) anomalies and curved southward, approaching the inner basin (Fig. 2.1.1, Shah et al., 2005). A

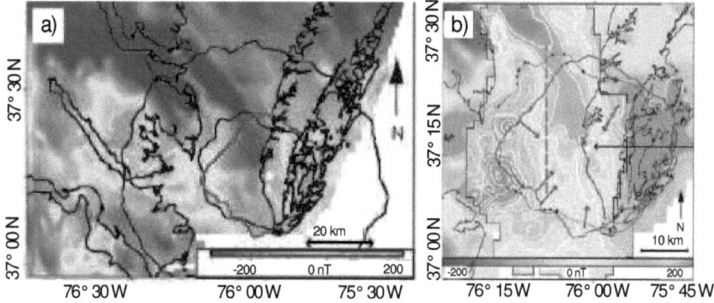

Figure 2.1.1: Magnetic field anomaly over Chesapeake Bay. a: The regional field anomaly is dominated by a magnetic low. Positive anomalies occur within the crater but extend beyond the crater rims. b: The central magnetic low is surrounded by several positive anomalies. Red arrows point to small circular positive anomalies. Contour interval 40 nT. Modified from Shah et al. (2005).

400 nT magnetic low appears above the crater's centre and is surrounded by a couple of positive anomalies to the east and west. Calculated magnetic gradients indicate that the numerous short-wavelength anomalies are attributed to shallow magnetic sources (Shah et al., 2005). Near and above the suggested central peak (Gohn et al., 2009b), steeply sloping high-low anomalies of~300 nT occur and Shah et al. (2009) explained this feature by an abrupt change of the basement lithology in this region (schist block on the E-side, gneiss block on the W-side). In combination with the data acquired by the Eyreville drilling in 2005 and 2006, Shah et al. (2009) developed a subsurface model which additionally includes crystalline megablocks that are intercalated in the sedimentary impact layers. The latter assumption best explains the long and short-wavelength magnetic anomalies. Furthermore, impact sediments and intercalated megablocks have been encountered by the drilling itself and are equally indicated by seismic observations (Gohn et al., 2008; Powars et al., 2009). Magnetic lows are generally attributed to shock-demagnetization, whereas magnetic highs are thought to result from melt formation or shock-induced

remagnetization (Pilkington and Grieve, 1992). On the basis of measured and estimated magnetic susceptibilitiy and natural remanent magnetization (NRM) values from the drilled lithologies Shah et al. (2009) attributed regions with a high magnetic gradient to shallow basement or melt blocks. After all, the measured magnetic field at the surface generally reflects the rock magnetic properties of the particular units beneath. The groundwork of a structural model that can be deduced from the measured magnetic field is therefore a data basis which provides the average magnetic quantities, like NRM and susceptibility, of each crater unit. In the following, the complete record of magnetic susceptibility and NRM measurements that has been collected from samples in the CBIS is presented. Data from Elbra et al. (2009) and Shah et al. (2009) will be used in the following along with own measurements of samples from the Eyreville, Cape Charles, Langley and Bayside cores (for locations see Fig. 1.1.2). The aforementioned works indicate that only suevite and crystalline basement contain the ability to carry a significant magnetization intensity. For this reason, our supplemental sampling procedure focused on these rock types.

2.1.2 Results

The different lithologies from the Eyreville core show a large variation of magnetic susceptibility (χ), natural remanent magnetization (NRM), and Königsberger ratio (Q). Fig. 2.1.2 gives an overview of all measured units obtained from the Eyreville drilling and their specific magnetic parameters. Corresponding average values are shown in Table 2.1.1. Impact breccia, sandstone, amphibolite, and pegmatite show average susceptibility values below $1 \cdot 10^{-3}$ SI, indicating the dominance of a paramagnetic magnetic mineralogy (Soffel, 1991). Thin sections of the Exmore breccia reveal hematite as a ferromagnetic component, but a low fraction of a magnetite-near phase is also described by Elbra et al. (2009) for this unit. It is, however, not known if these minerals are remnants derived from pre-impact units, or formed after shock. Minor amounts of magnetite are also mentioned for the granitic pegmatites (Elbra et al., 2009). The amphibolite blocks from the Eyreville and Cape Charles cores are both paramagnetic and contain small amounts of hematite. Their susceptibilities are rather low (in average $6.6 \cdot 10^{-4}$ and $3 \cdot 10^{-4}$ SI, respectively) and so is the NRM (2 and 3 mA/m, respectively). In contrast, granite, suevite, and schist, all deriving from the Eyreville core, are predominantly ferrimagnetic (Fig. 2.1.2). The granite has a noticeable average susceptibility of $1.6 \cdot 10^{-2}$ SI and NRM values of ~84 mA/m in average. This unit contains large magnetite grains with sizes of some tens of μm in average and is characterized by multidomain (MD) behaviour. The Q-values for this unit remain mostly below 1. Magnetite in this rock formed during the crystallization of the granite and is of a pre-impact origin. In the following, such magnetic phases will be named as "primary" minerals. The schist block contains pre-impact pyrrhotite that has undergone ductile deformation during a metamorphic event which equally produced the rock foliation. Brittle deformation has overprinted all grains in different intensities and is interpreted as brittle faulting associated with the crater modification stage after passage of the release wave (Horton et al., 2009a). Large scattering of susceptibility and NRM occurs in the schist, but maximum values of 3030 mA/m

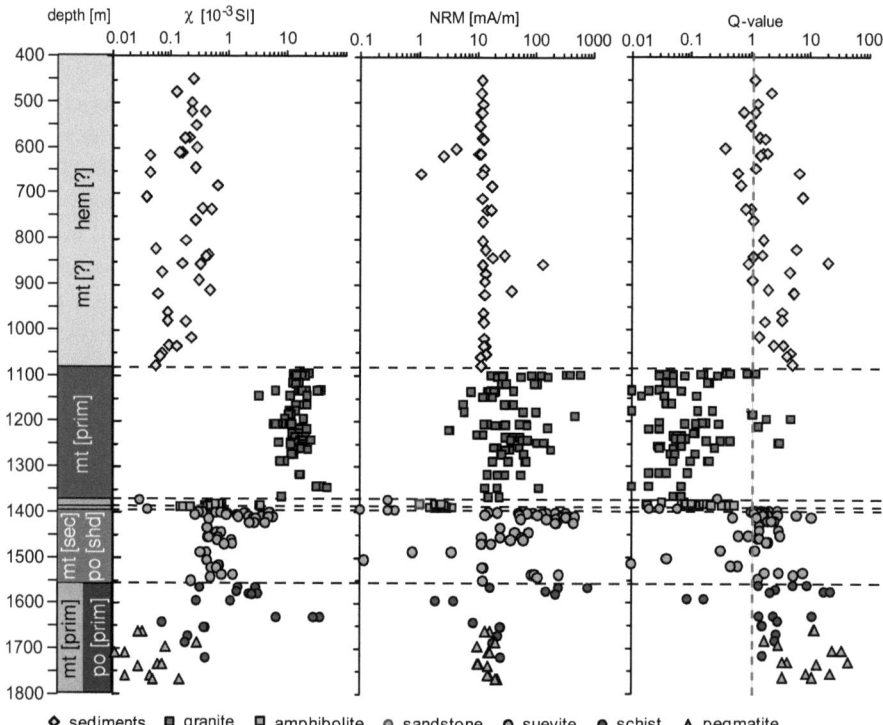

Figure 2.1.2: Rock magnetic properties of lithologies from the Eyreville core. Granite, suevite and schist show susceptibilities exceeding $1 \cdot 10^{-3}$ SI, but among these units only suevite and schist are able to carry a stable magnetization (Q-value > 1). NRM and susceptibility from sediments are too small to contribute to the magnetic anomalies. NRM: natural remanent magnetization; Q-value: Königsberger ratio: $\frac{NRM}{J_i}$ with J_i (induced magnetization) = $\chi \cdot H$, and H = 41 A/m); χ: Susceptibility.

occur (not shown in Fig. 2.1.2). The same holds for the susceptibility, which is $4.9 \cdot 10^{-3}$ SI in average, but comprises values up to $35 \cdot 10^{-3}$ SI. Scattering is also strong in the upper part of the suevite unit (1400 – 1422 m, Fig. 2.1.2). Compared to the suevite section below 1400 m, this part contains grains with a strong magnetization up to ~500 mA/m (Fig. 2.1.2) and also susceptibilities range up to $5 \cdot 10^{-3}$ SI. The raised susceptibilities indicate higher quantities of ferrimagnetic minerals in this upper suevite section. This section is also characterized by a high impact melt fraction (Wittmann et al., 2009b). The latter authors detected Fe-Ti-oxides within impact melt streaks which they interpreted to be magnetite. However, own studies revealed exclusively shocked pyrrhotite within melt streaks and fragments, but no indications of magnetite crystallized from the melt. In this study magnetite was found within the suevite matrix instead. These grains have been formed after deposition of the suevites and are part of the

Table 2.1.1: Average values and standard deviations of natural remanent magnetization (NRM), magnetic susceptibility (χ) and Königsberger ratio (Q) of the different rock types from the four drill cores in the Chesapeake Bay impact structure.

	NRM [mA/m]	st. dev.	χ [SI]	st.dev.	Q	n
Eyreville						
sediment	16.19	$1.96 \cdot 10^{-2}$	$2.06 \cdot 10^{-4}$	$1.49 \cdot 10^{-4}$	1.91	41
granite	83.84	$3.95 \cdot 10^{-1}$	$1.61 \cdot 10^{-2}$	$7.54 \cdot 10^{-3}$	0.13	88
sandstone	0.28	$1.26 \cdot 10^{-4}$	$1.78 \cdot 10^{-4}$	$1.65 \cdot 10^{-4}$	0.04	4
suevite	125.13	$1.35 \cdot 10^{-1}$	$1.60 \cdot 10^{-3}$	$1.59 \cdot 10^{-3}$	1.91	46
schist	754.26	$7.27 \cdot 10^{-1}$	$4.85 \cdot 10^{-3}$	$9.97 \cdot 10^{-3}$	3.79	18
pegmatite	16.96	$4.11 \cdot 10^{-3}$	$6.31 \cdot 10^{-5}$	$7.46 \cdot 10^{-5}$	6.56	15
amphibolite	2.38	$4.37 \cdot 10^{-4}$	$6.64 \cdot 10^{-4}$	$5.36 \cdot 10^{-4}$	0.09	39
Cape Charles						
amphibolite	2.74	$3.01 \cdot 10^{-4}$	$5.63 \cdot 10^{-4}$	$3.06 \cdot 10^{-4}$	0.12	15
gneiss	7.35	$7.45 \cdot 10^{-3}$	$1.55 \cdot 10^{-3}$	$1.76 \cdot 10^{-3}$	0.12	32
suevite	10.21	$1.16 \cdot 10^{-2}$	$5.22 \cdot 10^{-4}$	$6.66 \cdot 10^{-4}$	0.48	27
basalt	433.95	$2.14 \cdot 10^{-1}$	$1.57 \cdot 10^{-2}$	$1.54 \cdot 10^{-3}$	0.67	8
Bayside						
granite	44.54	$4.86 \cdot 10^{-2}$	$1.06 \cdot 10^{-2}$	$7.61 \cdot 10^{-4}$	0.1	37
Langley						
granite	31.04	$1.05 \cdot 10^{-1}$	$1.51 \cdot 10^{-3}$	$3.10 \cdot 10^{-3}$	0.5	41

n: number of samples; st. dev. standard deviation

secondary mineral fraction which formed during hydrothermal post-impact alteration. The upper suevite part is therefore simply characterized by a higher amount of magnetite grains that are likely the result of impact melt alteration (see chapter 2.2). Abundant lithic rock components are mainly of sedimentary origin and do not contain recognizable fractions of ferrimagnetic minerals. They are also absent in the few observed schist fragments with diameters of ~200 µm in average. However, Horton et al. (2009b) found yet unknown gneiss and granitoid fragments in the suevite unit and such rocks usually contain magnetite (e.g. Archanjo et al., 1995; Petrík and Broska, 2007; Trindade et al., 1999). However, these rocks will not be discussed in the following, since they were not observed in the studied samples. The suevites are dominated by a remanent magnetization, but also most samples of the schist block are. In contrast, remanence of the granite megablock is rather unstable and confirms the presence of MD magnetite (Soffel, 1991).

Compared to the Eyreville core, some differences in magnetic properties occur in the Cape Charles core (Fig. 2.1.3). Remanent magnetization of all units is usually below 100 mA/m and mostly unstable. In all units magnetite is the exclusive ferrimagnetic mineral, as confirmed by reflected light microscopy and SEM studies. The suevite contains predominantly shocked magnetite and grains strongly vary in size with a main fraction of about 2 - 30 µm. Variations in grain size also occur in the gneiss, but in this unit,

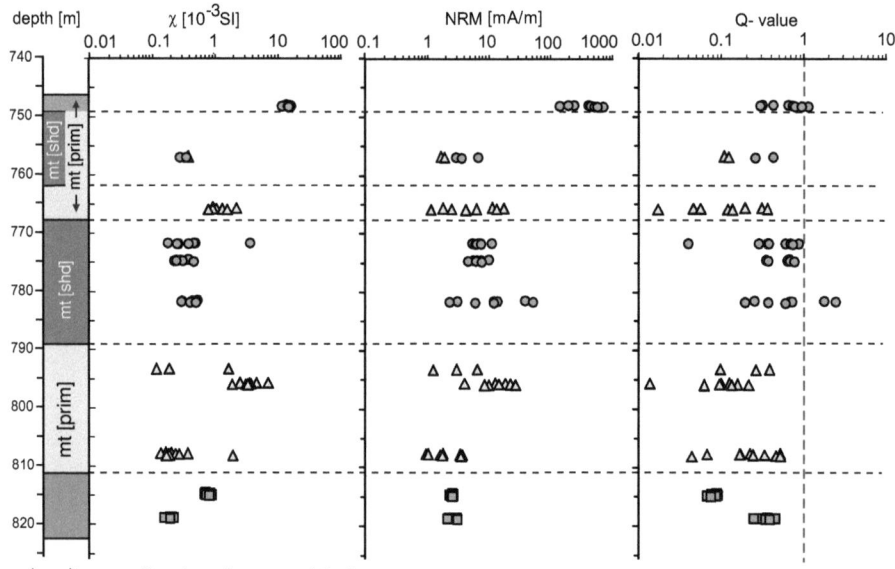

Figure 2.1.3: Rock magnetic properties of lithologies from the Cape Charles core. Only gneiss and basaltic dyke blocks exceed susceptibilities of $1 \cdot 10^{-3}$ SI, but induced magnetization is dominant in all samples.

the main fraction is several tens to hundreds μm in diameter. Parts of the gneiss are strongly altered which is illustrated by the presence of secondary chlorite and white mica. This rock shows the highest susceptibilities in the lower part between 815 and 820 m, where alteration is less strong than above. The gneiss unit is characterized by induced magnetization (Q-values below 1, Fig. 2.1.3). Ferrimagnetic minerals were not found in thin sections of the amphibolite block, which shows rather low susceptibilities (below $1 \cdot 10^{-3}$ SI) and remanent magnetizations (below 10 mA/m). Finally, the mafic dyke block shows the significantly highest NRM values (in average: 0.43 A/m) and remarkable susceptibilities of $1.6 \cdot 10^{-2}$ SI in average. But Q values mainly remain below 1, indicating that induced magnetization dominates over remanent magnetization. This rock contains dendritic titanomagnetite with diameters from below 0.1 μm up to 3 μm. Although magnetic susceptibilities are high ($15 \cdot 10^{-3}$ SI), NRM values are below 1 A/m. These NRM values are distinctly different to the approximation of 2 and 6 A/m for melt rocks reported in Shah et al. (2009, see Table 1 therein). Since these authors had no own data available, they based their approximation for the basaltic block, which they interpreted as impact melt, on data from the melt section of the Eyreville core. However, this rock lacks of any evidence indicating the presence of impact melt. Such evidence could be lithic clasts in the matrix or a turbulent fabric. Both features are common in impact rocks (Wittmann et al., 2009b). Instead, the rock holds a typical basaltic mineralogy, including olivine (up to 600 μm in size), plagioclase, and few μm-sized magnetite dendrites. The whole mineralogy is strongly altered and titanomagnetite therein is not able to carry a stable magnetization.

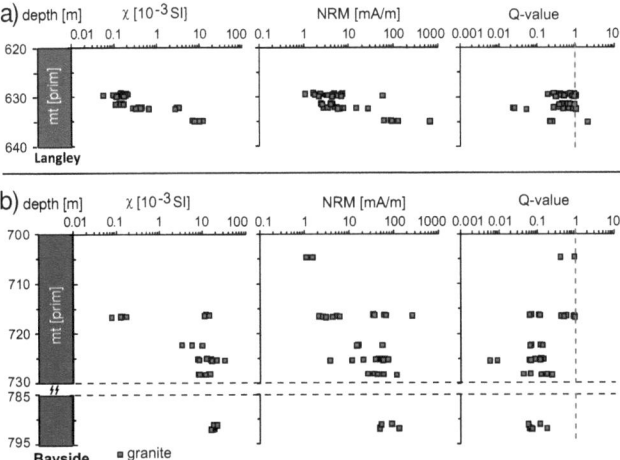

Figure 2.1.4: Rock magnetic properties of lithologies from the Langley and Bayside cores. a: Langley core: Susceptibilities of most samples are below $1 \cdot 10^{-3}$ SI and the granite samples carry no stable remanence. b: Bayside core. Susceptibilities of most samples lie above $1 \cdot 10^{-3}$ SI, but induced magnetization is dominant in that unit.

The Langley granite (Fig. 2.1.4a) is pervasively altered and alteration is suggested to have mainly occurred before the impact (Horton et al., 2005a). Most susceptibility values are below $1 \cdot 10^{-3}$ SI and NRM values remain below 100 mA/m, except of a few samples from the bottom part of the drilling. Under the optical microscope these samples appear less altered than the ones from further above. All Q-ratios are below 1, indicating that the induced magnetization dominates over the remanent magnetization. Most of the Bayside granite (Fig. 2.1.4b) samples are ferrimagnetic with an average χ value of $10 \cdot 10^{-3}$ SI, but NRM values are below 200 mA/m. The rock is weakly altered, but some stronger altered samples show distinctly lower χ and NRM values (717 m).

Demagnetization behaviour

A detailed study of the demagnetization behaviour of the Eyreville samples has already been presented by Elbra et al. (2009). Within this study some additional measurements were taken, especially from the suevite and schist units, in order to find clues for the magnetization mechanism during the impact process. A brief overview and interpretation of the bulk data set will be given in the following. Because the drill cores were not oriented, only inclination but no declination values were retrieved from alternating field (AF) and thermal demagnetization. For demonstration of the magnetic hardness, coercivity values (Table 2.1.2) will be used instead of mean destructive field (MDF) values. The latter do not provide a reliable data set, since large jumps in magnetization occurred during the demagnetization experiments in a broad sample fraction.

Granite samples from all boreholes contain MD magnetite which is magnetically soft (Table 2.1.2). Elbra et al. (2009) reported a main remanence component of the Eyreville granites with a shallow inclination (~10°), but most granite samples do not show a stable direction during thermal or AF demagnetisation. However, few shallow inclinations are also present in the Langley granite (Table 2.1.3) and they are in agreement with the interpretation of Elbra et al. (2009), who suggested that the Eyreville granite had preserved a pre-shock remanence which was present in the basement rocks. These authors found that the pre-shock remanence had shallow inclinations of ~10°. The Langley granite is located in the outer cater which is supposed to contain in situ and tilted basement megablocks. Its inclination therefore indicates that this rock either represents an in situ basement rock, or an only slightly tilted megablock. AF demagnetization of the granite samples reveals that low fields of ~15 - 20 mT are sufficient to almost completely demagnetize the samples and evidence a soft magnetic behaviour. The schist block does not carry a stable remanence vector, with the exception of one sample (Fig. 2.1.5B, Table 2.1.3) which has a steep inclination of 74° and an average coercivity of 29 mT. The remanence is carried by pyrrhotite and is removed at an unblocking temperature close to T_{Cpo} = 325°C. Within the suevite, the remanence behaviour is largely inconsistent and varies from sample to sample. Many samples from the Eyreville core show an unstable magnetic remanence vector, but some samples contain a stable component (Fig. 2.1.5C, D). However, these stable components show different inclinations compared to each other and therefore they are randomly oriented with depth. A relationship between orientation and depth is not recognizable. Basically, a set of shallow inclinations (1.6 - 25°) occurs together with a set of steep inclinations (70 - 80°, Table 2.1.3). Additionally, two moderately dipping inclinations were measured with values at 64.2 and 67.2°. The dip of inclination is independent of weather magnetite, or pyrrhotite and magnetite are present in the particular sample. In the latter case, about 30% of the magnetization is carried by pyrrhotite and both minerals share the same direction (Fig. 2.1.5D). The average coercivity value for the magnetite-bearing Eyreville suevite is ~22 mT, regardless of the fact if pyrrhotite is present or not (Table 2.1.2). Samples from the Cape Charles core all show an unstable remanence vector (Fig. 2.1.5E) and the average coercivity values are around 16 mT indicating a softer magnetic behaviour compared to the Eyreville suevite. The basalt carries a stable

Table 2.1.2: Average coercivity (H_c) values of the particular impact units.

unit	n	mineralogy	H_c [mT]
Eyr granite	3	mt prim	0.5
CC basalt	1	mt prim	25.5
Eyr schist	3	po prim	29.1
Eyr suev (a)	7	mt sce+po shd	23.4
Eyr suev (b)	5	mt sec	22.1
Eyr suev all	12	mt sec+po shd	22.8
CC suev	5	mt shd	15.9

Eyr: Eyreville; CC: Cape Charles; suev: suevite; mt: magnetite; po: pyrrhotite; prim: primary; sec: secondary; shd: shocked.

direction with an inclination of ~20° and has a coercivity of 26 mT.

2.1.3 Discussion

Rock magnetic properties of the different lithological units in the CBIS depend on the particular magnetic carriers and their pre- and post-impact evolution. Mechanical and chemical changes can modify the type of remanence and the specific magnetic properties (Dunlop and Özdemir, 1997). Post-impact alteration below 100°C is described to have pervasively affected the suevite (Wittmann et al., 2009a, b), and less pervasively the crystalline units, but some of the latter rocks have experienced their main alteration before the impact (Horton et al., 2005, 2009b). Concerning the rock magnetic properties in this study, the time of alteration is not important if the remanence has not been modified by some impact-related process. Instead, the initial remanence and the present degree of alteration are arbitrative. Among others, alteration is evident for the three main granite units, gneiss and the basaltic dyke. Except the gneiss, all these units carry a thermoremanent magnetization (TRM) which they acquired during crystallization from an initial melt. The particular Q-values show that magnetite in all these units is not able to acquire a remanent magnetization. All granites contain MD magnetite, but show different susceptibility and NRM values.

Table 2.1.3: Magnetic lineation values of the studied samples

sample	depth [m]	rock type	incl. [°]	demag type
L1-2 (L)	634.9	granite	12.6	therm
L1-4 (L)	625	granite	5.1	therm
L1-8 (L)	635.1	granite	10.1	AF
B1-7 (B)	725.4	granite	34.8	therm
CC 4-1 (CC)	795.6	gneiss	-45.8	therm
CC 13 (CC)	748	basalt	19.7	AF
CB 20 (Eyr)	1399	suevite	-14.1	therm
CB 21 (Eyr)	1400.1	suevite	1.6	therm
CB 22 (Eyr)	1400.5	suevite	64.2	AF
CB 23 (Eyr)	1401	suevite	21.8	therm
CB 24 (Eyr)	1401.2	suevite	13.5	AF
CB 25 (Eyr)	1404.6	suevite	-23.1	AF
CB 26 (Eyr)	1408.6	suevite	3.5	therm
CB 26 (Eyr)	1408.6	suevite	74.9	AF
CB 27 (Eyr)	1409.1	suevite	67.2	AF
CB 29 (Eyr)	1421.7	suevite	5.1	therm
CB 33 (Eyr)	1576.4	schist	74	AF

L: Langley; B: Bayside; therm: thermal; AF: alternating field; incl.: inclination; demag: demagnetization.

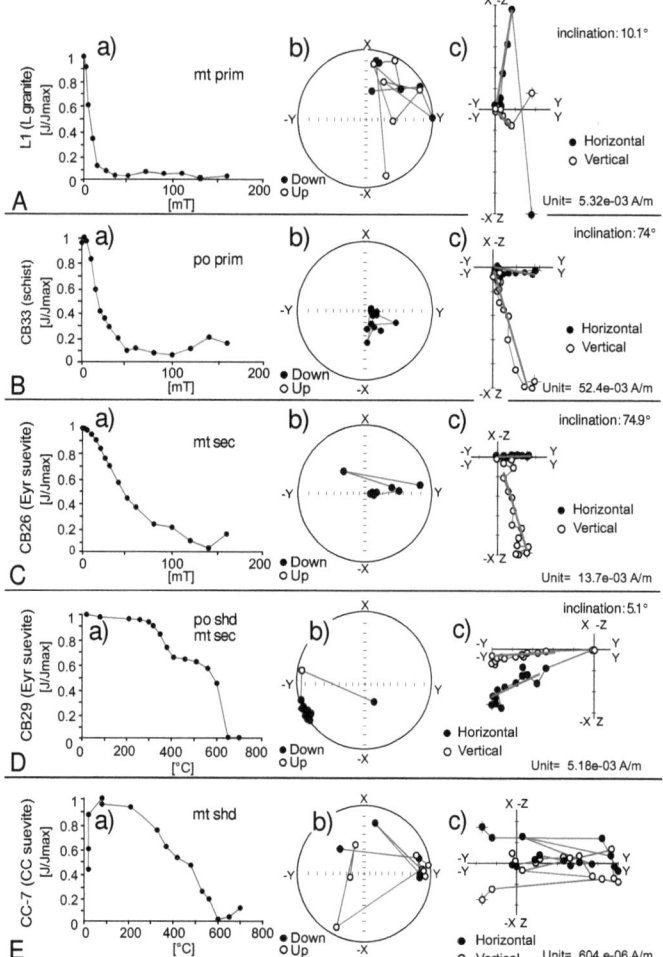

Figure 2.1.5: Examples of demagnetization experiments. A - C: Alternating field (AF) demagnetization. a: Intensity decay (J/Jmax) versus magnetic field curves, b: Stereographic projections of remanence directions, c: Orthogonal (Zijderveld) vector projections. D - F: Thermal demagnetization. D: After heating up to~350° the magnetization of pyrrhotite is removed, but magnetite is still present. No change in direction of remanence vector occurs above that temperature, indicating that both minerals carry the same direction. a: Intensity decay (J/Jmax) versus temperature curves, b - c: equal projections as in A - C.

For comparison, all data have been plotted in discrimination diagrams (Fig. 2.1.6). These diagrams allow a clear distinction of two magnetic different granite types. Microscopic studies revealed that the main difference is the degree of alteration. Granite with high susceptibilities (e.g. from the Eyreville core and partially from the Bayside core) contains magnetite, while the most altered granite (Langley granite) shows the lowest susceptibilities (Fig. 2.1.6c). Low susceptibilities are related to the oxidation reaction from magnetite to hematite. The magnetic susceptibility of hematite is low ($\chi = 1.2 \cdot 10^{-6}$ SI along the c-axis, Dunlop and Özdemir, 1997) and its formation at the expense of magnetite therefore lowers the bulk susceptibility and intensity of the NRM. Magnetite from the basaltic block is also pervasively altered, wherefore its relatively low NRM values lie below 1 A/m. The grain sizes of titanomagnetite from this rock suggest that a single domain (SD) fraction (below 0.1 μm) is present which, in return, should be able to carry a stable remanent magnetization (Dunlop and Özdemir, 1997). A pre-impact remanent magnetization is also present in the schist block, but a stable direction is mostly missing. Since pyrrhotite usually belongs to the metamorphic rock assemblage (e.g. Rochette, 1987) it contains a CRM or a TRM, depending on the temperatures during metamorphosis. Gibson et al. (2009) suggested peak-temperatures of 600 - 670 °C which is distinctly above the blocking temperature ($T_{CPo} = 320°C$) of pyrrhotite. It is therefore likely that this rock carries a TRM. This TRM, however, is not stable, since the large pyrrhotite grains are dominated by MD behaviour. The mostly unstable demagnetization behaviour of the schist rock is in accordance with observations of Elbra et al. (2009), who were not able to isolate a stable component during demagnetization of samples from this unit. On the one hand, the general unstable demagnetization behaviour indicates that the steep inclination of the schist sample CB 33 (Fig. 2.1.5C, Table 2.1.3) represents a drilling induced magnetization. On the other hand, the Q-factors of these rocks mostly lie above 1. Therefore, this rock should not easily acquire a new magnetization. The initial magnetization has probably been destroyed during alteration which has largely affected the schist rock and pyrrhotite therein (see chapter 1.1 and 2.3). In Fig. 2.1.6, the Eyreville suevite can be quite well distinguished from the Cape Charles suevite. Whereas the Eyreville suevite contains shocked pyrrhotite and secondary magnetite, shocked magnetite was found as the only magnetic carrier in the Cape Charles suevite. Secondary magnetite should carry a CRM which was acquired after the impact. Either the shallow or the steep inclination could represent the magnetic field of that time. Since shallow inclinations occur in shocked pyrrhotite, this may indicate that such directions are representative for the CRM acquired after the impact. Although it is not yet clear if pyrrhotite carries a CRM or a shock remanent magnetization (SRM), the inclination is the same in either case, since both remanence types would have been acquired subsequently after the impact. At first, it therefore seems reasonable that magnetic remanence directions of shocked pyrrhotite and secondary magnetite share the same orientation. However, data from several intrusions between 42 and 49 Ma suggest paleomagnetic inclinations between -55 and -61° (Lovlie and Opdyke, 1974; Ressetar and Martin, 1980) and similar values (-56°) were present between 6 and 8 Ma (Heinrichs, 1967). Likely, magnetic inclinations did not significantly change between both periods and the suevite-related inclinations can therefore not directly be linked with the magnetic field at the time of the impact. The only exception is the occurrence of the two moderate inclinations (Table 2.1.3) in secondary magnetite-bearing samples. These inclinations are in accordance with the proposed inclinations during the Eocene and could therefore indicate a post-impact CRM. However, the unstable magnetic be-

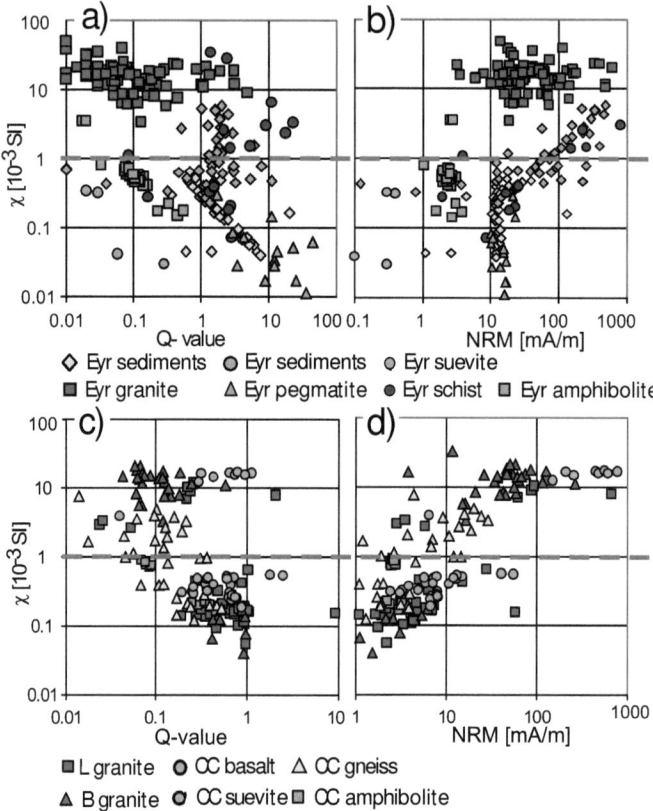

Figure 2.1.6: Discrimination diagrams for all data. a: Susceptibility vs. Q-value (Eyreville core). Most suevite samples exceed Q-values of 1 and the granite shows susceptibility values mainly above $10 \cdot 10^{-3}$ SI in accordance with a low degree of alteration: this granite is termed as "fresh". b: Susceptibility vs. NRM. Large scattering of NRM is visible in the suevites. c: Susceptibility vs. Q-value (Cape Charles, Langley and Bayside core). Langley and Bayside granites can be clearly distinguished. Whereas samples below $1 \cdot 10^{-3}$ SI indicate advanced alteration, the "fresh" samples scatter around $10 \cdot 10^{-3}$ SI. The former are mostly related to the Langley, and the latter to the Bayside granites. In contrast to the Eyreville suevites, the Cape Charles suevites have Q-values below 1, indicating the dominance of induced magnetization. d: Susceptibility vs. NRM. Only the basalt block exceeds NRM values of 100 mA/m.

haviour of many suevite samples during thermal demagnetization indicates that the present phases can relatively easily be remagnetized and the steep inclinations thus argue for a drilling induced magnetization. Louzada et al. (2010) have shown that shock is not able to completely demagnetize pyrrhotite above 3 GPa. In the experiments of these authors, the remaining remanence increased with pressure. Thus, one may argue that the shallow inclinations indicate remanences in pyrrhotite that have survived the impact process. However, shallow inclinations are also present in samples which contain exclusively magnetite. Furthermore, orientation of the pyrrhotite grains was randomized during the ejecta deposition. This is different to the assumption Elbra et al. (2009) made on the deposition of some crystalline megablocks. Although both units contain similar inclinations, no logical link between both can thus be found in the framework of this thesis, concerning the origin of their particular remanence. Shallow inclinations in the suevite are not fully understood and may result from non-uniform magnetic directions in the suevite. Further discussion on this topic can be found in the study of Elbra et al. (2009).

Regarding shocked magnetite from the Cape Charles core, it appears that these grains did neither acquire a stable uniform remanence after the impact, nor during drilling. This is similar to a large fraction of shocked pyrrhotite grains. The results of this study therefore show that unstable remanence directions occur throughout large parts of the crater. The unstable magnetic behaviour of the suevite indicates that neither magnetite nor pyrrhotite acquired a relevant magnetization, and if they did, this magnetization was very weak and unstable. However, coercivities between 16 and 29 mT for the suevites (Table 2.1.2) suggest that an acquired remanence should be not easily demagnetized, but in fact conserve a stable remanence. This observation is different to findings from the Bosumtwi crater, Ghana, where pyrrhotite is also the main magnetic carrier, but stable remanence directions in accordance with the age of the impact structure are abundant (Kontny et al., 2007).

2.1.4 Conclusions

The CBIS contains a large range of yet discovered crater units which are highly variable with respect to their rock magnetic properties. Ferrimagnetic behaviour is prevalent in the Eyreville and Bayside granites, but also in the upper part of the Eyreville suevite and the schist block. A significant remanent magnetization above 100 mA/m was only measured in suevite, schist, and some granite samples. Maximum NRM values are always below 4000 mA/m. Among these units only suevite and schist are able to carry a stable NRM, since granite is characterized by an induced magnetization. All granites and the basaltic dyke contain magnetite with a primary TRM. The unstable demagnetization behaviour of most samples, however, indicates that large fractions of the TRM have disappeared, likely due to alteration. The few stable remanence directions show shallow inclinations ($\sim 10^\circ$), which is in agreement with the 250 Ma old North American basement (Elbra et al., 2009). Alteration is strongest in the Langley granite and significantly reduces the ferrimagnetic behaviour of this rock. Bayside and Eyreville granites are rather fresh and some samples of both units exceed NRM values of 100 mA/m. Investigations on the "melt rock" can neither confirm the origin as a melt rock nor its high NRM (< 0.9 A/m), as proposed by Shah et al. (2009). Microscopic investigations indicate this unit to represent a mafic dyke, which is in agreement with its rock magnetic properties. The schist block contains exclusively pyrrhotite which is suggested to carry the initial remanence. Since pyrrhotite usually forms during metamorphic events and temperatures during metamorphosis of the schist rock largely exceeded the blocking temperature of pyrrhotite (320 °C), this rock likely carries TRM. However, remanence directions in most schist samples are unstable and the stable one is either drilling-induced or represents an initial magnetization. If the magnetization is of pre-impact origin, the megablock has been presumably tilted during deposition, since no steep inclinations are reported for the in situ basement rocks of Chesapeake Bay (Elbra et al., 2009). The suevite units show a strong variation in their magnetic properties, which is related to the abundance of the magnetic carriers: magnetite and pyrrhotite. Whereas shocked pyrrhotite and secondary magnetite control the magnetic properties of the Eyreville suevite, shocked magnetite is the dominant magnetic mineral in the Cape Charles suevite. Mostly the magnetic remanence vector is unstable during demagnetization, but some stable steep and shallow inclinations occur in the Eyreville suevite. Steep inclinations can be explained by the acquisition of a borehole-induced magnetization, but the shallow inclinations are hard to interpret, since dips of inclinations at the time of the impact were presumably about -55 to -60°. Although a link between shallow inclinations in the basement units and similar inclinations in the suevite is ruled out, due to random orientation of magnetic minerals in the suevite, it is remarkable that these inclinations occur simultaneously in basement units and impact sediments. After all, these results indicate that SRM acquisition played an insignificant role in the CBIS, since this process did not affect a measurable fraction of shocked magnetic minerals. The results of this study provide a significant database on the rock magnetic properties of the known impact units of the CBIS, if combined with the results of Elbra et al. (2009). This database allows specifying the model of Shah et al. (2009) and also confirms their basic assumption. According to the latter, the short-wavelength features in the magnetic field measurements of Shah et al. (2009) are related to basement-derived megablocks embedded in lithic and suevite breccia. Nevertheless, the details of the model need some modification:

- no melt rock with average NRM higher than 1 A/m has been observed

- lithic and suevite breccia show strong magnetic contrasts depending on the source rocks (schist, gneiss, amphibolite, granite, pegmatite) and the occurrence of magnetic minerals in suevitic matrix; therefore they can not be assumed as a homogeneous layer

- granite, schist and "melt rock" contribute with induced magnetization values of about 1 A/m to the total magnetization

The results of these studies suggest that the direct and indirect consequences of shock on rock magnetic properties of impact lithologies are rather complex and result in an interplay of various processes. Besides larger scaled shock features like brecciation and randomization of previous orientations, shock-induced modification of the particular magnetic carriers plays an important role. Especially since pre- and post-impact alteration affects the shocked grains, the shock-related modifications in the magnetic properties are not easy to determine. Such an approach requires a thorough study of the particular magnetic carriers in order to understand their specific magnetic behaviour. Magnetic properties, microstructures and the particular geological setting allow an estimation of the remanence origin and its contribution to the bulk magnetization of a specific unit. In the following chapters magnetite and pyrrhotite will therefore be intensively characterized by means of rock magnetic, compositional, and microscopic studies. A combination of microstructural and magnetic methods with respect to shock deformation in magnetite and pyrrhotite is unique up to now, but it will be shown that this approach is indispensable, if one wants to understand the complex interaction of shock and alteration-induced structural and magnetic features.

2.2 Microstructures and magnetic properties of pre- and post-shock magnetite in the CBIS

Abstract

This section focusses on the different generations of magnetite occurring throughout all lithological units of the CBIS. Based on their relation to the impact event, three main types are defined. Primary magnetite belongs to the magmatic mineral assemblage of granite and basaltic dyke and was therefore formed before the impact event. This type has not been notably affected by shock and shows a typical multidomain behaviour. Shocked magnetite was also formed before the impact event, but has been modified by shock. The most obvious and important shock-related deformation feature is grain size reduction and this feature is a product of shock-induced fracturing. Some grains are partially molten and contain degassing holes. Two different Verwey transition temperatures (T_V) can be detected in shocked magnetite, and this observation can be used for a distinction of three subtypes. Subtype I contains exclusively a lowered T_V (LT_V) around 85 K, subtype II shows a regular T_V at 123 K and mixed magnetite assemblages (subtype III) contain both subtypes. Secondary magnetite was formed after the impact during post-impact hydrothermal activity. This type forms mesh-like structures overgrowing the suevite matrix, within which various nm- to mm- sized single grains are intergrown. The single grains mostly show acicular grain shapes, which were presumably inherited from a goethite (FeOOH) precursor. Goethite can transform directly or via hematite to magnetite (e.g. Özdemir and Dunlop, 1993). Additional to a superparamagnetic and single domain fraction (subtype I) many clusters contain MD grains resulting in a pseudo-single-domain (PSD) behaviour (subtype II). Temperature-dependent susceptibility curves of subtype I lack of a T_V and indicate that the Curie transition occurs over a temperature interval of 100°C from about 540 to 640°. Subtype II shows a lowered (LT_V: 95 – 100 K) or broadened T_V in the same curves and the Curie transition sharply drops at 580°C. Investigations of this study have shown that LT_V occurring in shocked and secondary magnetite is mainly a result of oxidation. The crucial factor controlling the decrease of T_V is the surface/volume ratio of the particular grains. Oxidation of small grains in both magnetite types has been sufficient to affect this transition. However, different mechanisms are responsible for the formation of small magnetite particles. Whereas shocked magnetite is a product of mechanical fracturing, small grain sizes of secondary magnetite are entirely growth-related.

2.2.1 Introduction

The magnetic and structural transitions of magnetite (Fe_3O_4) are not only helpful for relatively quick identification, but also bear fundamental information on stoichiometry and the magnetic domain state

(Dunlop and Özdemir, 1997). These transitions are therefore commonly analyzed during low and high temperature (LT and HT, respectively) measurements in order to characterize magnetite-bearing rocks (e.g. Carpozen et al., 2006; Jackson et al., 1998; Liu et al., 2003). Besides the Curie temperature (T_C) at 580°C, the important transition in magnetite is the Verwey transition (T_V), at which the cubic structure is transformed into a monoclinic symmetry (e.g. Honig, 1995; Verwey, 1939; Williams, 1953). The transition is associated with an isotropic point (T_i) where the magnetocrystalline anisotropy constant becomes zero at 130 K (e.g. Halgedahl and Jarrad, 1995; Özdemir et al., 1993). Changes of the magnetic behaviour at T_V are highly sensitive to grain and magnetic domain size (Carter-Stigliz et al., 2006; Özdemir and Dunlop, 1999), stoichiometry (Aragón et al., 1985; Shepherd et al., 1985), internal stress (Ramasesha et al., 1994; Todo et al., 2001) and the magnetocrystalline anisotropy (Kosterov, 2001; Muxworthy and McClelland, 2000). T_V is present in all but superparamagnetic (SP) domain states. The most crucial factors influencing this transition are oxidation and the presence of impurities (Moskowitz et al., 1998; Özdemir and Dunlop, 1993). Whereas small deviations from ideal stoichiometry affect T_V but not T_i (Kąkol and Honig, 1989), higher degrees of non-stoichiometry depress and lower both transitions (Özdemir et al., 1993). The degree of decrease and suppression is dependent on the oxidation rate. However, Carpozen et al. (2006) found two T_Vs in shocked rocks of the Vredefort impact structure, South Africa. Besides the regular T_V, a second lowered T_V could be observed in this study. The authors correlated both transitions to different grain size fractions. In doing so, they associated the lowered transition with a smaller grain size fraction occurring within amorphous lamellae (planar deformation features, PDF) of shocked quartz. On the basis of the argument that the quartz host provided a relatively good protection against alteration, these workers excluded oxidation to be the cause of the lowered T_V.

The most common way to analyze T_V and T_i is the appliance of LT measurements. LT behaviour is usually analysed by imparting a saturational isothermal remanent magnetization (SIRM) on the sample, which is measured during a cooling and warming cycle from 300 K to 10 K and back. The magnetic behaviour around T_V, however, is critically dependent on the temperature (usually 10 or 300 K) at which the SIRM has been given, if cooling occurs in the presence of a magnetic field (Carter-Stigliz et al., 2006; Kosterov, 2001). In RTSIRM curves, the strongest loss of remanence occurs in true multidomain (MD) grains and decreases linearly with grain size (Halgedahl and Jarrad, 1995). Single domain (SD) grains with shape anisotropy do not undergo a T_V as their anisotropy is determined by particle shape (Dunlop and Özdemir, 1997), but uniform SD particles do show the T_V (Özdemir et al., 1993).

Threshold sizes with respect to specific domain states in magnetite define maximum particle sizes, below which a new domain state occurs. In general, the magnetic domain state is associated to a characteristic magnetic behaviour. The commonly accepted SD threshold ratio for magnetite, below which SD behaviour occurs, is 0.1 µm (Williams and Dunlop, 1989). This is true for cubic particles, but increasing crystal axis ratios enlarge the critical size length up to 1 µm (Dunlop and Özdemir, 1997). Pseudo-single-domain (PSD) behaviour is said to occur in grains with average diameters between 1 and 10 µm, whereas larger grains are truly MD (Butler, 1998). The threshold size of SP behaviour is not clearly defined but appears to lie around 0.01 μm (Goya et al., 2003; Rogwiller and Kündig, 1973).

2.2.2 Results

Since the crater fill of the CBIS consists of units which are very inhomogeneous in origin and composition (see chapter 2.1), different varieties of magnetite are present in those rocks. Whereas some types mainly reflect the initial pre-shock state, others have been modified by shock or formed as secondary minerals within the impact sediments. On the basis of their particular relation to the impact event, three main types of magnetite are defined in this study. Firstly, primary magnetite represents grains which have not been noticeably affected by shock and mainly preserved their pre-impact properties. Secondly, shocked magnetite comprises grains which have suffered minor or higher shock treatment, implying at least strong fracturing. Thirdly, secondary magnetite refers to grains which entirely formed after the impact and do not have a pre-impact history. General features of all three magnetite types are described in the following, using characteristic examples for each particular unit.

Types of magnetite

Primary magnetite (mt prim) occurs in the basaltic block and all sampled granites comprising the Eyreville (Fig. 2.2.1a), Bayside and Langley granites. Besides some differences in their pre-impact alteration state (Horton et al., 2006, 2009a + b), magnetite shows similar behaviour in all granite units. T_V and T_C are well-developed and appear as sharp and large changes in susceptibility in the χ-T curves (Fig. 2.2.1b) suggesting a stoichiometric composition. The latter show a reversible behaviour upon warming and cooling for the unaltered samples, whereas altered samples denote a more irreversible curve progression. Intergrowth with sphene and ilmenite are common. The grains are ususally some hundreds μm in size and are thus clearly MD. Magnetite in the basaltic dyke, in contrast, consists of dendrites ranging between 0.1 and 3 μm, indicating a mainly PSD-like behaviour.

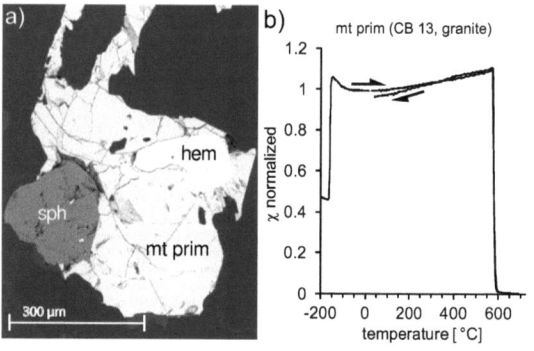

Figure 2.2.1: Primary magnetite (mt prim). a: large grains of magnetite intergrown with sphene (sph) and hematite (hem) from the Eyreville granite, BSE image. b: χ-T curve of the granite sample shown in a) measured in argon atmosphere. Heating and cooling cycle to 700°C are completely reversible. This curve is typical for stoichiometric magnetite without any significant signs of alteration. BSE: backscattered electron image; χ: susceptibility.

The whole rock, inclusively magnetite, has been strongly altered within the pre-impact setting (Horton et al., 2008). The alteration of magnetite has mainly produced hematite in all these rocks.

Shocked magnetite (mt shd) appears within the Cape Charles suevite and in the gneiss blocks or fragments. The gneiss has suffered higher shock pressures (Horton et al., 2009a) and magnetite therein is strongly fractured, but larger grains are still present. The initial grain size mostly comprised some hundred μm, but the abundant fractures produced present grain sizes between few to some tens of μm (Fig. 2.2.2a). Alteration has affected the grains before, but also after the impact (Horton et al., 2009a)

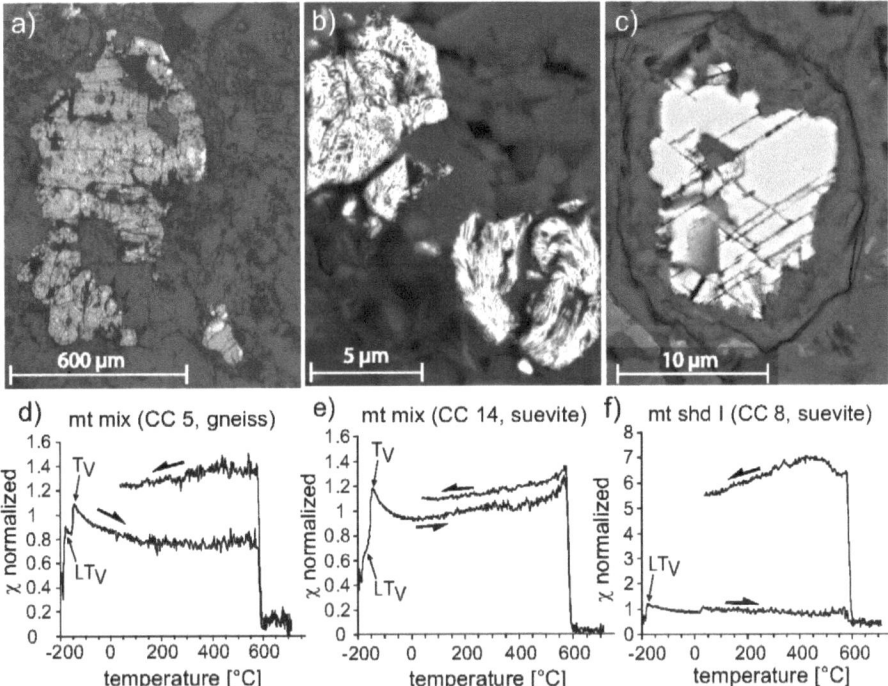

Figure 2.2.2: Shocked magnetite (mt shd). a: Strongly fractured magnetite from gneiss block, coated with ferrofluid (reddish colour), OM image. Microfractures with a preferred orientation (horizontal) indicate shock overprint. Strong alteration is visible at cracks and grain boundaries. b: partly molten grains with structures resembling turbulent flow patterns, BSE image. c: single magnetite grain from the Cape Charles suevite with two sets of lattice preferred cracks, which are similar to planar fractures (PF). Sutured grain boundaries indicate dissolution, BSE image. d: χ-T curve of gneiss sample shown in a) measured in argon atmosphere. Besides the regular Verwey transition at -153°C (T_V) a second susceptibility jump appears at −173 (LT_V) indicating the additional presence of a "modified" magnetite population (both T_Vs: mt mix). Additional magnetite has formed above T_C, BSE image. e: χ-T curve of a Cape Charles suevite measured in argon atmosphere showing T_V and LT_V (mt mix). Compared to mt mix in d), the LT_V is slightly broadened in this sample. f: χ-T curve of a suevite sample containing shocked magnetite I (LT_V) measured in argon atmosphere. The LT_V appears at -173°C and the total amount of magnetite is rather low in the heating curve. During cooling a significantly higher amount of magnetite is present, indicating reactions under reducing conditions that result in the transformation of hematite, Fe_2O_3, into magnetite. OM: optical microscope.

and is distinctly progressed in all observed samples. Magnetite grains in the suevite are generally present as single grains enclosed in the cataclastic matrix and range from 2 to 30 µm. Intensive penetration by fracturing or heating is typical (Fig. 2.2.2a + c) and evidences shock-triggered modifications. Fractures usually occur in one or two preferred orientations (Fig. 2.2.2c) and are similar to Planar Fractures (PFs). Heated particles show a turbulent fabric indicating partial melting (Fig. 2.2.2b). Such grains also contain abundant circular holes that presumably formed from degassing during shock heating.

χ-T curves of shocked magnetite are mainly irreversible and indicate the formation of magnetite during heating above 600°C (Fig. 2.2.2c). Such behaviour is well known for crustal rocks containing altered magnetite or hematite, which are reduced to magnetite in the argon atmosphere of the sample holder (Just and Kontny, 2012). The additional fraction of magnetite upon cooling explains the higher susceptibilities in these runs. Within the different samples, two distinct T_Vs can be observed: one at -178°C (95 K) and another at -153°C (120 K) (Fig. 2.2.2d - f). The first T_V appears at a distinctly lower temperature than the well known, second T_V and will be named low-T_V (LT_V) in the following. LT_V is present in all suevite samples, whereas T_V appears only in some samples of that unit. This feature, in return, is contrariwise in the gneiss samples. The appearance of LT_V indicates the presence of a somehow modified fraction of magnetite grains. On the basis of this assumption, a subdivision of shocked magnetite will be undertaken as follows: Samples showing exclusively a LT_V are referred to as subtype I (mt shd I), whereas samples with a regular transition are denoted as subtype II (mt shd II). Samples containing both subtypes are described as mixed magnetite assemblages (mt mix).

Secondary magnetite (mt sec) occurs in the suevite from the Eyreville core and forms porous grain clusters (Fig. 2.2.3a - c). These clusters consist of plenty nm- to µm-sized, randomly oriented magnetite grains, which are grown together in a mesh-like structure. Two types of single grains occur within a cluster (Fig. 2.2.3c). Type 1 consists of uniaxial monocrystals with length of 1 to 10 µm and diameters up to 0.3 µm. The high aspect ratio indicates a needle-shaped or platy habitus. However, no plane-like grains are recognizable from the microscopic photographs, wherefore a rod shape is the more likely interpretation. Type 2 consists of nm-sized xenomorphous crystallites forming polycrystalline domains. Grain sizes generally range over a broad interval, but the upper size limit is not consistent. Some clusters contain distinctly larger single grains while others are limited to a lower size. All grains occur within the frame of a cluster and cannot be found as outstanding components in the suevite matrix. The clusters themselves are free of any visible chemical alteration, and selected area diffraction (SAED) patterns obtained by TEM exclusively display magnetite reflexes. The clusters overgrow older cataclastic structures (Fig. 2.2.3a) indicating that the time of their formation is younger than the deposition of the suevite, which implies a post-impact origin. Similar to shocked magnetite, a LT_V appears between -180 and -170°C (~90 - 100 K) in these samples. Unlike to shocked magnetite, the transition is often blurred and develops continuously between LT_V and T_V. Some samples lack completely of a T_V. In few samples the susceptibility drop at T_C is strongly blurred and smeared out over a temperature interval of up to 100°C, from about 540 to 640°C (Fig. 2.2.3e). Heating is often accompanied by oxidation above ~450°C, indicated by a constant rise of susceptibility above this temperature. Such behaviour shows that the present magnetite phase is not stable and is characteristic for strongly oxidized or very fine grains. Typically, curves showing a lower T_C (Fig. 2.2.3e) are associated with a missing T_V, whereas curves that display a T_V show T_Cs at 580°C or even exceed this temperature due to the formation of an oxidized heating artefact.

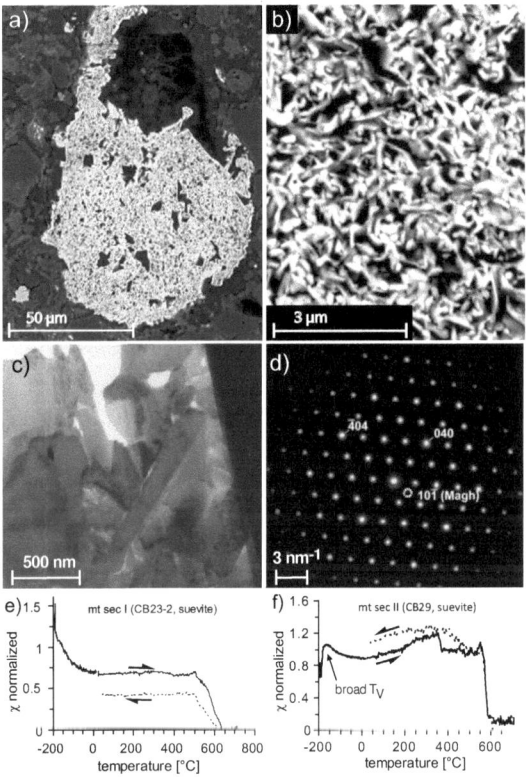

Figure 2.2.3: Secondary magnetite (mt sec) from the Eyreville suevite. a + b: Mesh-like magnetite aggregates consist of differently oriented needle-shaped single crystals, BSE image. c: TEM image showing the size of these single crystals. A large monocrystal (lower right hand side) is surrounded by various nanocrystals. d: SAED pattern of the monocrystal in c. Only magnetite reflexes occur, the position of the 101 maghemite reflex (Magh) is marked for reference. e + f: χ-T curves of Eyreville suevite. While some curves show no T_V and T_Cs starting around 540°C (e) (mt sec I), others show a LT_V around -180°C and a regular T_C at 580°C (f) (mt sec II). Both curves were measured in argon atmosphere and are irreversible. TEM: transmissions electron microscope; SAED: selected area electron diffraction pattern.

Hysteresis parameters

The so called Day plot (after Day et al., 1977) is generally used to classify magnetic domain states with the help of the main hysteresis parameters (M_{rs}/M_s vs. H_{cr}/H_c). Using this plot gives a relatively good discrimination of the three magnetite main types from the CBIS (Fig. 2.2.4). Average values for the relevant parameters are given in Table 2.2.1. Low M_{rs}/M_s (0.0038) and very high H_{cr}/H_c ratios (30) are characteristic for the granites, wherefore these samples plot in the MD field. Subtype II of shocked magnetite, which occurs in the gneiss sample, has very similar values for the main parameters (M_{rs}/M_s = 0.0052, H_{cr}/H_c = 51) and plots near the granites and far to the right from all other samples. Both primary magnetite and subtype II of shocked magnetite are excluded from Fig. 2.2.4 in order to give a more detailed perspective on the smaller-sized samples plotting between SD and MD field. Dendritic magnetite from the basaltic dyke plots on a SD-MD mixing line between the PSD and SD field, which is in accordance with the SD-MD grain sizes known from microscopic observations (0.1 - 3 µm) in this section. The SD-MD mixing line has been calculated by Dunlop (2002) for magnetite in the PSD range, and this trend line fits approximately well with the

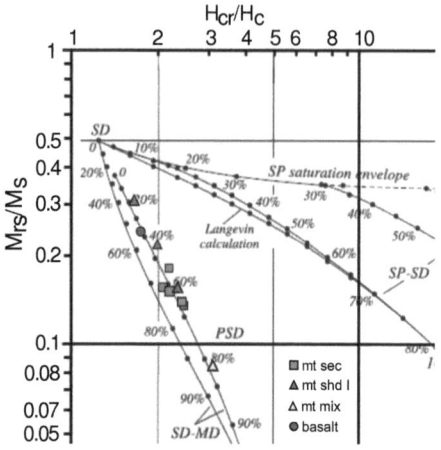

Figure 2.2.4: Day plot (after Day et al., 1977) with theoretical mixing curves calculated from Dunlop (2002). Plots from samples from this study appear close to the SD-MD mixing line. Whereas the grain size distribution in shocked magnetite is strongly variable, secondary magnetite clusters around a M_{rs}/M_s ratio of 0.15. The relatively large fraction of SP grains is not recognisable in the mixing curve. Pure MD grains (primary magnetite) are not shown and occur in the MD field, far below the illustrated section. SD: single domain; MD: multidomain; M_s: saturation magnetization; M_{rs}: remanent saturation magentization; H_c: coercivity; H_{cr}: remanence of coercivity.

Table 2.2.1: average hysteresis parameters for the three main types of magnetite in the CBIS

	M_{rs}/M_s	B_{cr}/B_c	M_{rs} [Am²/kg]	M_s [Am²/kg]	H_c [mT]	H_{cr} [mT]	n
mt sec	0.3209	1.55	0.0572	0.177	29.01	44.49	5
st. dev.	(746)	(039)	(420)	(124)	(785)	(1065)	
mt shd I	0.2305	2.04	0.0054	0.021	22.55	44.91	3
st. dev.	(629)	(023)	(039)	(001)	(471)	(430)	
mt shd II	0.0052	50.69	0.0018	0.343	0.72	36.62	1
mt mix	0.8052	3.13	0.0039	0.046	11.18	35	1
mt prim	0.0038	30.34	0.0030	1.04	0.54	13.81	3
st. dev	(017)	(1418)	(010)	(793)	(026)	(264)	

st.dev.: standard deviation; n: number; mt shd I: LT_V; mt shd II: T_V; mt mix: $LT_V + T_V$.

three samples of shocked magnetite (subtype I). These samples are parallel to, but typically slightly above the SD-MD mixing line and are in very good agreement with the observed grain sizes ranging from <1 µm to several µm. Since these small grain sizes are mainly a product of shock - induced by brittle deformation, the sample position on the mixing line approximately reflects the fracturing intensity. The sample of mixed magnetite plots on the same trend line but towards lower M_{rs}/M_s and higher H_{cr}/H_c values, indicating a larger fraction of MD grain sizes. Curiously, secondary magnetite clusters around the mean values of M_{rs}/M_s = 0.194 and H_{cr}/H_c = 2.312, which is in agreement with the data of Day et al. (1977) for PSD grains but contrary to what one would expect from the microscopic studies. Despite the fact that some single crystals have length of few µm, one would expect a dominant SD behaviour to arise from the clusters as the high aspect ratio should enhance the SD threshold (Dunlop and Özdemir, 1997). SD-dominated grains, however should give M_{rs}/M_s = 0.5 (Dunlop, 2002), which is far above the ratio for secondary magnetite. Furthermore, the large quantity of nm-sized crystallites should generate a superparamagnetic (SP) behaviour, which is said to shift the plotted data points towards the SD-SP mixing curve in Fig. 2.2.4.

First-order reversal curves (FORC)

The unexpected hysteresis parameters of secondary magnetite give rise to the question if magnetic interactions occur within the particular clusters and hence lower the M_{rs}/M_s ratio (Dunlop and Özdemir, 1997). Therefore, FORC measurements were conducted on samples which exclusively contain secondary magnetite. The obtained FORC diagrams are helpful in characterizing domain state and determining magnetic interactions between magnetic grain assemblages.

Fig. 2.2.5 shows two diagrams for secondary magnetite that cover the range of grain sizes in these samples. A large vertical peak at very low coercivity (H_c) close to the origin of the diagram arises from thermal relaxation effects (e.g. Roberts et al., 2000) and is typical for SP behaviour (Fig. 2.2.5a). Behind this peak, the contour shape becomes narrow and elongated and a second peak arises around 15 mT. The low vertical spreading and the close- to symmetric shape around the main peak suggest a fraction of SD grains with rather low magnetic interactions between the single grains. However, small interactions are indicated by the slightly asymmetric contour shape below the zero axis (Roberts et al., 2000), but such effects are negligible in this plot.

In Fig. 2.2.5b, a peak at the origin is also present but less well-developed compared to Fig. 2.2.5a. This may be grounded in a lower fraction of SP grains, but the relatively large smoothing factor (SF = 5) additionally modifies the development of this peak. The shape of the main peak is distinctly larger along the H_u axis ($H_u = \frac{H_a+H_b}{2}$, with H_a: reversal field of the FORC and H_b: any point on the FORC, see also Roberts et al. (2000)) and has an asymmetric shape in relation to the H_c-axis. The main outline is relatively compressed and values for H_u from the outer contour do not exceed 60 mT, which finally indicates a grain size fraction in the PSD range (Muxworthy and Dunlop, 2002; Roberts et al., 2000).

Similar shapes make up most of the FORC diagrams, whereas SP-SD shapes appear less frequently. The magnetite clusters obviously contain a wide grain size distribution ranging from SP nm-sized crystallites to MD μm-sized monocrystals, which is in agreement with the microscopic observations.

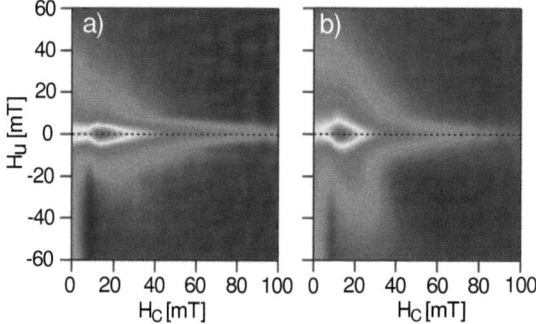

Figure 2.2.5: FORC distribution of two selected samples containing secondary magnetite from the Eyreville suevite. Both samples contain a first peak at very low H_c values, which is sparsely developed in these diagrams due to the relatively high smoothing factor of SF equal to 6. A fraction of superparamagnetic low-coercive grains can, however, be recognized by the strongly deflecting contour shape next to the H_u axis. a: the flat and almost symmetric distribution indicates SD grains, which are almost free of magnetic interactions. b: The asymmetric contour shape with a distinct waist below zero indicate PSD behaviour. The total distribution is noteworthy since it suggests the appearance of SP, together with SD and MD grains. SP: Superparamagnetic; for explanation of H_u see text.

Low temperature behaviour

SIRM given at room temperature (RTSIRM) was measured for one sample from Eyreville granite (primary magnetite, CB 13), two samples from the Cape Charles suevite (shocked magnetite I, CC 7, and mixed magnetite, CC 14) and one sample from the Eyreville suevite (secondary magnetite, CB 23). Fig. 2.2.6 shows the obtained measurement curves. The curve for primary magnetite is typical for that of MD magnetite (e.g. Özdemir and Dunlop, 1999; Özdemir et al., 2002) showing a remanence drop on cooling that reaches its maximum, at which the remanence is 23% of initial value, at 121 K. When T_V is passed, SIRM partially recovers and a small second drop appears at about 85 K. Below that point SIRM remains temperature-independent and completely reversible. During subsequent warming T_V appears at 126 K and remanence drops to a minimum of 18% of the initial remanence. This drop, however, is not permanent as the remanence fully recovers on further warming with respect to its value at 10 K and ends with a final remanence of 31 % of the initial value.

The curves of primary and secondary magnetite are almost identical in shape and do not exhibit a remanence recovery when passing T_V. On cooling a significant remanence decrease appears at T_V. Below that point, remanence is temperature-independent and completely irreversible. However, compared to primary magnetite, the remanence drop for secondary magnetite is smaller, and the SIRM memory is

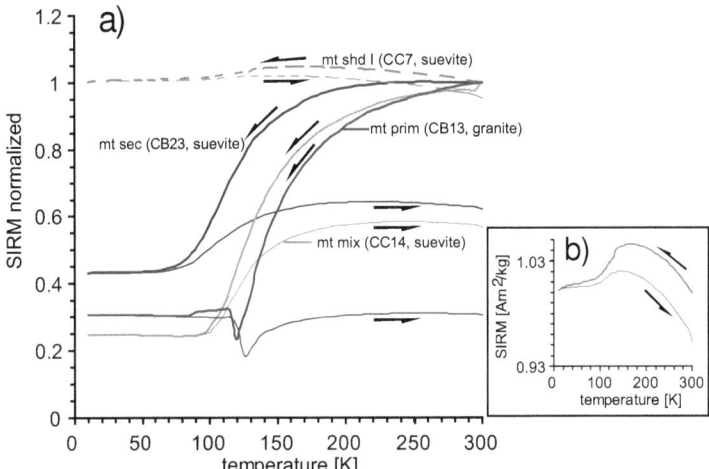

Figure 2.2.6: Saturation isothermal remanent magnetization (SIRM) behaviour during cycling from RT to 10 K and back to RT. a: Primary magnetite (mt prim) showing a typical MD behaviour. The remanence strongly drops at cooling, but partially recovers below T_V. Subsequent heating beyond T_V causes the remanence to fall to a minimum but fully recovers with further warming. SIRM of secondary magnetite (mt sec) drops until ~80 K and contains 42% of the initial remanence, indicating that the transition is partly suppressed. The mixed magnetite assemblage (mt mix) shows a T_V around 95 K and strong loss of remanence (76%) at this point. Only LT_V can be deduced from this curve. Shocked magnetite (mt shd) has lost ~5% of its initial remanence after total cycling and the curves remain comparatively flat. Curves of this sample are enlarged in b). b: Strong temperature dependence above 120 K suggests instable magnetic domains. Furthermore, the transition at T_V is hampered and remanence remains irreversible down to 10 K, which can be most likely ascribed to structural defects in the lattice structure.

slightly higher (62% for mt sec and 57% for mt prim). Furthermore, the constant part of the curve of secondary magnetite is reached at a lower temperature, between 70 and 80 K. Shocked magnetite, in contrast (enlarged in Fig. 2.2.6b), shows an increase in SIRM up to a maximum at about 160 K upon cooling and a subsequent drop until 10 K, where the SIRM memory is approximately equal with the initial value. The SIRM memory after warming to 300 K is about 95%. Both SIRM memory and curve shape show that T_V and T_i are strongly suppressed in this sample.

Field cooled (FC) and zero field cooled (ZFC) remanence curves were measured for secondary magnetite and mixed magnetite and the curves of these samples differ in shape (Fig. 2.2.7). For both samples, the SIRM of the ZFC curve is higher than that of the FC curve. This behaviour is typical for PSD-MD particles (e.g. Carter-Stiglitz et al., 2006). During warming, SIRM of mixed magnetite first decreases moderately until a distinct drop occurs at 90 - 95 K. Subsequently, an even sharper drop appears between 120 and 123 K. Although these features are poorly developed, the onset of the first drop at 95 K can be linked to the LT_V appearing in the χ-T curve of this sample. Accordingly, the steeper remanence decrease around 120 K is associated with T_V in Fig. 2.2.2e. Secondary magnetite, in contrast, shows

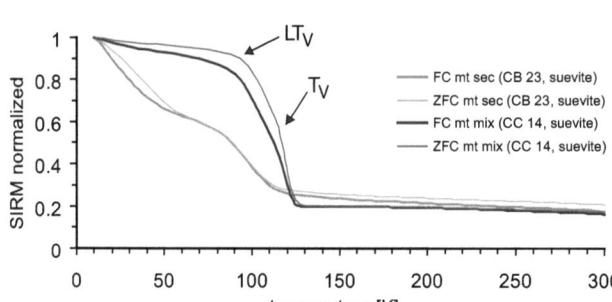

Figure 2.2.7: Field cooled (FC), respectively zero field cooled (ZFC) low temperature behaviour of secondary magnetite (mt sec) and mixed magnetite assemblages (mt mix). Quick drop of SIRM above 10 K in the curves of mt sec is indicative of SP behaviour. The decrease in remanence mainly ends at 110 K, but the curve remains temperature-dependent until 300 K. SIRM of mt mix shows two poorly developed transitions around 95 K (LT_V) and 120 K (T_V).

a stronger decrease in FC and ZFC remanence near 10 K and a T_V at 110 K, which is slightly lower than the one of mixed magnetite. However, the remanence decrease associated with T_V is not sharp, but occurs over a broad temperature range between 85 and 110 K. Furthermore, an additional kink appears at about 40 K. The drop at T_V is somewhat deeper for mixed magnetite and remains constant until RT, whereas remanence of secondary magnetite shows a slight temperature-dependence until 300 K.

2.2.3 Discussion

2.2.3.1 Primary and shocked magnetite

The remanence curves of primary magnetite (mt prim, Fig. 2.2.6 + 2.2.7) are in good agreement with MD behaviour of magnetite as described in the literature (e.g. Halgedahl and Jarrad, 1995; Muxworthy et al., 2003; Özdemir et al., 2002). In general, the remanence drop in RTSIRM curves is a general effect of progressive domain wall unpinning. At the same time, the remanence rebound below T_V in the curve of primary magnetite can be explained by the formation (on cooling) or break up (on warming) of twin structures in the monoclinic phase (Özdemir et al., 2002). These twins are formed as a result of internal strain induced by the phase transition at which the c-axis changes in length (Halgedahl and Jarrad, 1995). This effect is absent in smaller grains since such grains remain presumably untwined below T_V, because domain wall movement is inhibited by the smaller particle size.

The exclusive presence of a regular T_V, which defines subtype II of shocked magnetite, is restricted to some of the gneiss samples. These samples are consistent with MD behaviour which can in deed be expected from the observed grain sizes. The low coercivity (0.72 Am^2/kg, see Table 2.2.1) is similar to that of primary magnetite in the Eyreville granite (Ø: 0.54 Am^2/kg). This equally holds for the M_{rs}/M_s ratio, which is 0.0038 for granite and 0.0052 for gneiss. Brittle deformation has diminished grain sizes in gneiss (Fig. 2.2.2a), but in general the magnetic properties remain rather unaffected. However, the LT_V

at~95 K occurring in some gneiss samples (mixed magnetite, Fig. 2.2.2d) is worth mentioning since this temperature indicates a second fraction of magnetite which has been somehow modified. Unfortunately, no hysteresis and LT data are available for these samples, but similar LT_Vs appear in shocked and mixed magnetite from the Cape Charles suevites. In the following, the Verwey transitions and the hysteresis parameters of these shocked magnetites will be discussed. It is assumed that the results are equally valid for mixed magnetite in gneiss samples.

It has been shown that all samples of shocked magnetite lie on the SD-PSD mixing line (Fig. 2.2.4), but some contain larger fractions of MD grains than others. This observation is in agreement with the general strong inhomogeneities of grain sizes and shock features in the suevite, which are, in return, typical for impact sediments (e.g. Abadian, 1972; Horton et al., 2009a; Melosh and Ivanov, 1999). The largest fraction of MD grains occurs in mixed magnetite (CC 14), since the M_{rs}/M_s ratio (Table 2.2.1) is comparatively low, but also the remanence loss at T_V is large (Fig. 2.2.6a). In other words, the suevite sample containing the largest fraction of MD grains is the only sample of this unit holding a regular T_V in addition to the prevalent LT_V. It seems therefore obvious that the regular T_V of all samples arises from stoichiometric magnetite mainly ranging in the MD size.

In the following, LT_V will be discussed in more detail since it is present in a large fraction of samples that have somehow been modified by shock. To obtain a better overview, Fig. 2.2.8 summarizes the three main types of transition features associated with shocked and primary magnetite. Considering the specific properties, the focus will first be put on the RTSIRM curve of shocked magnetite, which exclusively contains a LT_V. The curve shape initially indicates titanomagnetite as a possible candidate (Özdemir and Dunlop, 2003). Titanomagnetite would also explain the lower transition temperature at LT_V (Moskowitz et al., 1998). However, microprobe and EDX analysis on various samples confirm that there is no T_i in these magnetites. The analyses also show that other impurities are negligible as well, at least in the studied grains. Even if titanomagnetite occurs within unmeasured magnetite assemblage, the temperature of T_C being at 580°C argues against the presence of such chemistry, because T_C of titanomagnetite appears at distinctly lower temperatures (e.g, Lattard et al., 2006; Moskowitz, 1981).

A LT_V in association with shock has already been described for magnetite from the Vredefort impact crater (LT_V denoted as LT-T_V, Carpozen et al., 2006), but the authors were not able to fully explain this phenomenon. On the basis of their study they defined a smaller- and larger-sized grain fraction. Whereas the former is suggested to have formed during or after the impact, the latter is of pre-impact origin. Both T_Vs could be clearly detected in all kinds of remanence and susceptibility measurements, which is different to the results of this thesis. In the here presented measurements, two values for T_V lying around 95 K (LT_V) and 120 K (T_V) occur in the χ-T curve of mixed magnetite (Fig. 2.2.2d, e + 2.2.8), but RTSIRM indicates only the LT_V, since this curve shows a rather constant decrease and becomes invariant below~95 K. The FC/ZFC curves weakly indicate the presence of two values for T_V at 95 and 120 K, but the development of LT_V is distinctly less than in the study of Carpozen et al. (2006). The latter

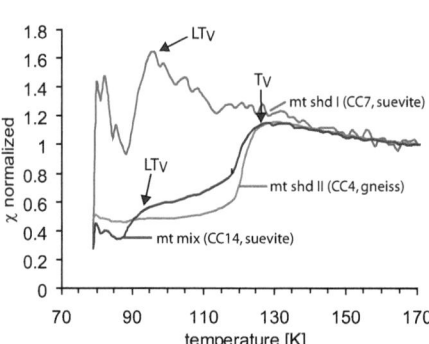

Figure 2.2.8: Susceptibility behaviour of three different shocked magnetite samples between 78 and 170 K measured with the kappabridge KLY-4A (AGICO) in argon atmosphere. All samples are normalized to their particular susceptibility at 170 K. The samples contain primary magnetite, shocked magnetite or both. Peaks between 78 K and 85 K are either attributed to an unknown phase or associated with a measurement artefact occurring instantly after the measurement was started. The latter interpretation is supported by the fact that the peak occurs only in sample CC 7, which has the lowest susceptibility (absolute χ at 170 K: CC 7: 6.7 SI, CC 14: 16.3 SI, CC 4: 67.5 SI). The flattening has levelled off above 85 K. Further changes in susceptibility are controlled by the magnetic minerals in the sample. Subtype I of shocked magnetite (mt shd I) shows one distinct susceptibility jump at 123 K (T_V), whereas this transition appears already at ~95 K (LT_V) for subtype II (mt shd II). The Susceptibility of the mixed magnetite assemblage (mt mix) increases in two steps, indicating the presence of both transitions.

study showed that the LT-T_V was associated with the fine grains and regular T_V with larger grains. Since the small-sized grain fraction is suggested to have formed during or after shock, these grains should not contain abundant shock deformation features. Shock-induced modifications of magnetite are therefore not a valid argument to explain the lowered T_Vs of these magnetite grains. Carpozen et al. (2006) excluded alteration as a cause for T_V lowering since the quartz host grain was argued to provide adequate protection against oxidation. This scenario is again different in our case. Alteration processes occurred pervasively after shock, especially in the suevite section (Horton et al. 2009b; Wittmann et al., 2009a) and are equally visible in thin sections. Magnetite in suevite and gneiss has been therefore distinctly affected by alteration, which, in return, usually implies oxidation. During oxidation of magnetite, vacancies are introduced into the crystal lattice and induce slight deviations from the regular stoichiometry (Aragón et al., 1985). Since T_V is very sensitive to non-stoichiometry (Aragón et al., 1985; Özdemir et al., 1993), LT_V is presumably a result of post-shock oxidation, especially for what concerns the small-sized magnetite fraction. Small grains are generally more susceptible to alteration processes because of their large surface/volume ratio. This interpretation is supported by the fact that both T_V and T_i are suppressed in the remanence curves of shocked magnetite, indicating a higher degree of non-stoichiometry (O'Reilly, 1984). Following this conclusion, shock deformation features are mainly responsible for large grain size reduction, but do not have direct effects on T_V. This interpretation agrees with the observation from Carpozen et al. (2006) in the fact that their LT-T_V was equally produced from unshocked magnetite. On the other hand, direct shock deformation effects de facto strongly modified shocked magnetite and should have created abundant defects within the crystal lattice. This becomes obvious when studying the thin sections or SEM photographs (Fig. 2.2.2a - c). Therefore, the consequent question is the following: If oxidation has such a strong effect on T_V, why do the abundant defects not affect T_V? The remanence behaviour of shocked magnetite (Fig. 2.2.6b) could indeed indicate the influence of defects since magnetic domains are not stable when cooled down below 10 K. During reheating, in contrast, the remanence remains rather temperature-independent until~80 K. The slight and continuous remanence drop below 80 K suggests that the transformation to the monoclinic phase is hampered and continues when cooling down below the main drop. Defects are the most likely explanation for this behaviour, since such microstructures influence the magnetocrystalline anisotropy and, more importantly, magnetostriction (Kosterov, 2001) by lowering the original symmetry. This happens in a similar way then for titanomagnetite after introduction of Ti^{4+} ions especially at low temperatures (O'Reilly, 1984; Özdemir and Dunlop, 2003). The fluttering cooling curve of Fig. 2.2.6b indicates that domain walls are not stable, not even in the monoclinic structure. However, defects usually pin magnetic domain walls and should therefore harden the magnetic behaviour (Dunlop and Özdemir, 1997). The observed behaviour is therefore likely result of interacting processes including temperature-dependent changes in magnetostriction and structural ordering as well as nucleation, pinning and movements of magnetic domain walls. Defects then weaken the domain wall stability and largely suppress T_V.

Finally, the influence of shock deformation structures is hard to estimate since the main cause for LT_V is the deviation from stoichiometry. Since oxidation and shock-induced defects are able to cause non-stoichiometry, both features are difficult to distinguish from the magnetic properties only. Oxidation is actually sufficient to explain the main magnetic features and therefore the role of the defects is hard to estimate. In fact, defects are not considered as an explanation for the lowered T_Vs in the study of Carpozen et al. (2006). In any case, the two transitions in mixed magnetite of the here presented study strongly indicate that oxidation has a distinctly stronger impact on small grains, whereas the increased surface/volume ratio of larger grains seems to leave T_V rather unaffected. This interpretation is confirmed by results from Özdemir et al. (1993) who found a relationship between the grain size and T_V for oxidized magnetite. In case of the CBIS, oxidation and shock deformation evidently affected magnetite, whereas shock deformation and especially grain size reduction dominantly acted as precursor for subsequent oxidation.

2.2.3.2 Secondary magnetite

Formation of needle-shaped magnetite

Idiomorphic magnetite typically forms octahedral crystals, which are a consequence of its cubic crystal symmetry. Elongated magnetite particles are therefore conspicuous and occur in nature mainly as product of magnetotactic bacteria (e.g. Linford et al., 2005; Mann et al., 1987; Matsunaga, 1991; Moskowitz, 1988). Such magnetite particles are usually organized in chains and have distinctly lower aspect ratios than clustered magnetite of this study. An organic origin seems therefore rather unlikely. Since the single grains are randomly oriented and the needles are not connected to cracks or veins, competition growth can be equally ruled out to explain the high aspect ratio. It therefore appears that the present grain shape is inherited by a precursor mineral. Needle-shaped magnetite has indeed been synthesized in the last decades by artificially induced polymorph transformations (Lian et al., 2003; Penn et al., 2006; Wang et al., 1998). All such synthesis procedures have in common that goethite (α-FeOOH) was used as needle-shaped precursor and was then transformed into magnetite by controlled changes of the redox conditions. During these reactions, magnetite was formed as a polymorph and adapted the crystal shape of the original goethite. A similar process has been described by Kobayashi (1959) and Özdemir and Dunlop (1993) for natural environments. They describe the transformation from goethite or lepidocrocite (γ-FeOOH, equally needle-shaped) to hematite (α-Fe_2O_3). Under reducing conditions, hematite is, in return, easily transformed to magnetite (Haigh, 1957). The initial grain shape is maintained during all these reactions. Fig. 2.2.9 shows phase relations in the Fe-O-H system and suggests that a transformation from goethite to magnetite is also possible as direct reaction. Following these relations, goethite can be transformed into magnetite by either reducing the Eh- (path 1) or the pH-conditions (path 2). Both reactions depend on the initial redox conditions and are presumably triggered by modification of both parameters when occurring under natural conditions.

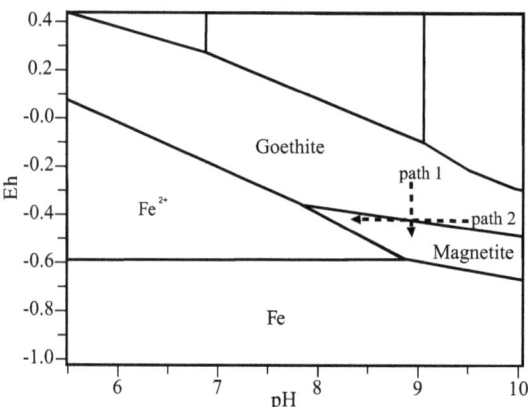

Figure 2.2.9: Phase relations of the Fe-O-H system at standard conditions (T = 273.15 K, p = 1 bar). Two theoretical paths are shown, which allow the transformation from goethite (FeOOH) to magnetite (Fe_3O_4). Path 1 requires reduction of the Eh value, whereas path 2 results in the reduction of the pH value. Under natural conditions such a transformation is presumably performed by changing both parameters. Modified from Neff et al. (2005).

These transformations obviously require a distinct change of the chemical conditions. Such a scenario is indeed realistic when applied to the suevites from the CBIS. After deposition of the impact lithologies, hydrothermal activity is usually triggered by the strongly heated central crater in complex craters (e.g. Melosh and Ivanov, 1999; Wittmann et al., 2009a). In the case of the CBIS, large amounts of seawater were involved since the impacted area was located at the continental shelf (Gohn et al., 2004). Hydrothermal activity has therefore induced chemical reactions, especially in the rocks near the central crater. As seen in chapter 2.1, magnetite is more abundant in the upper suevite section, which is characterized by higher amounts of impact melt. However, crystallized magnetite could not be found in the melt fragments. Instead, abundant secondary clusters occur near those fragments. Apparently, the precursor iron oxide was, at least partially, formed from products released during decomposition of impact melt. In general, melt particles are thermodynamically unstable (Marshall, 1961) and quickly affected by alteration (Vernaz, 2001). Since the composition of suevites is in general strongly inhomogeneous, alteration-induced changes of the chemical conditions seem very likely. Organic matter, for example, can distinctly reduce the redox conditions and allow the reduction of Fe^{3+} to Fe^{2+}, which is needed when hematite or goethite are transformed into magnetite. Such reductive material can be formed for example by graphite-bearing schist fragments that occur within the suevite section of the Eyreville core (Bartosova et al. 2009; Wittmann et al. 2009a). Depending on the rate of the particular mineral decomposition and the resulting fluid compositions, the chemical environment in impact sediments can undergo distinct changes (Hecht et al., 2004; Newsom et al., 1986; Stähle, 1972). Such alteration processes most likely allowed the formation of secondary magnetite in the CBIS suevites.

Magnetic properties

As described above, a large range of grain sizes is present in the secondary magnetite clusters from the Eyreville suevite. Each grain size fraction gives a specific magnetic signal, which again modifies the bulk signal. If one fraction dominates the magnetite assemblage, the magnetic signal will mainly reflect the properties of this particular fraction. This observation is important for what concerns the remanence measurements. SP, SD, and PSD behaviour occurs in the magnetite clusters, but the appearance of the associated grain fractions in susceptibility and remanence measurements differs within the samples.

The total size of most needle-shaped monocrystals is too large for a true SD behaviour. Even though shape anisotropy raises the boundary for SD behaviour, the M_{rs}/M_s ratio is much too low to reflect a large fraction of such grains (Özdemir et al., 2002). Predictions based on modelling of ferrimagnetic domain structures (Williams and Dunlop, 1989) link the average M_{rs}/M_s ratio of 0.155 (Table 2.2.1) to grain sizes of ~0.2 μm. This estimation is in good agreement with the TEM observations for the monocrystals, but not sufficient for pure SD behaviour, which has a threshold particle size of 0.1μm. Moreover, these parameters suggest that monocrystals must form both SD and MD grains, mainly depending on the length of their long axis. Since the range of stable SD magnetite is very small (~0.01 – 0.1 μm, Dunlop and Özdemir, 1997; Goya et al., 2003; Rogwiller and Kündig, 1973) and the grain size distribution in the clusters is rather large, the grain size fraction coinciding with SD behaviour must be comparatively low. This interpretation is consistent with the RTSIRM measurements of these grains (Fig. 2.2.6), indicating a typical PSD assemblage (e.g. Halgedahl and Jarrad, 1995). It also agrees with the M_{rs}/M_s and H_{cr}/H_c ratios of these samples, which cluster around the MD-PSD trend line in far distance to the SP-SD trend line in the Day plot (Fig. 2.2.4). In addition, the ZFC curve of secondary magnetite is elevated above the FC curve which is typical for PSD behaviour (e.g. Kosterov, 2003). Finally, PSD-like assemblages in most of the samples are indicated by FORC distributions. Hence, a large fraction of single grains possess grain sizes with multiple magnetic domains. This explains the presence of Verwey transitions in most samples.

The results of this study showed that the development of T_V is strongly variable within the sam-

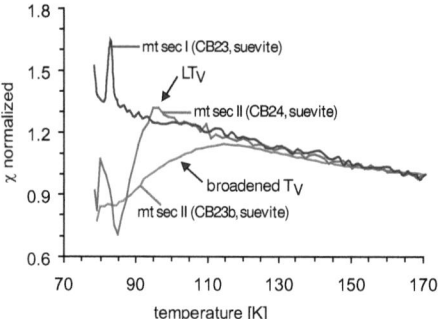

Figure 2.2.10: Susceptibility behaviour of three secondary magnetite samples between 78 and 170 K measured with the kappabridge KLY-4A (AGICO) in argon atmosphere. All samples are normalized to their particular susceptibility at 170 K. Interpretation of the peaks between 78 and 85 K is analogous to that given in Fig. 2.2.8 (absolute χ at 170 K: CB 23b: 41 SI, CB 24: 7.5 SI, CB 23: 5.85 SI). Whereas some clusters do not show a T_V at all (mt sec I), other clusters show a quite clear transition (mt sec II). In the latter case the transition appears at LT_V or is broadened. This results in a successive increase in susceptibility between LT_V and 115 - 120 K, which is close to the normal T_V.

ples containing secondary magnetite (Fig. 2.2.10). These variations allow one to draw several conclusions which will be developed in the following. The strong decrease of the FC/ZFC curves (Fig. 2.2.7) just above 10 K is evident for very low unblocking temperatures and strongly indicates SP magnetite (Özdemir et al., 1993). The presence of such a fraction is confirmed by the FORC diagrams. Spontaneous reversals occur in SP particles within a very short time range (Dunlop and Özdemir, 1997) and lead to a complete demagnetization in a zero field. Since SIRM is completely lost in the first minutes of the experiment, SP particles show no T_V (Moskowitz et al., 1989; Özdemir et al., 1993; Pardoe et al., 2001; Rogwiller and Kündig, 1973). A missing T_V is also ascribed to SD grains with particle shape (Dunlop and Özdemir, 1997). It has been shown that such grains are present in form of needles ranging below the SD threshold ratio. Both grain size fractions should therefore not be the cause of the observed LT_V. The counter-argument is that clusters producing χ-T curves without T_V (Fig. 2.2.3e + 2.2.10) consist of single grains that lie in the SP-SD range. A similar conclusion can be drawn from the FORC diagram in Fig. 2.2.5a. The contour shape suggests the exclusive presence of a SP and SD grain size fraction and leads to the conclusion that magnetite clusters in this sample must mainly consist of small grains below 0.1 µm. A further indicator for a limited grain size within such clusters is the early onset of T_C (Fig. 2.2.3e), which is strictly associated with a missing T_V. Cao et al. (2007) developed a model of ferrimagnetic nanoparticles predicting that T_C is predominantly dependent on grain size rather than on particle shape for grain sizes in the nm-scale. Within these conditions, the authors predict a decreasing T_C with decreasing grain size. Applying this model to our results, the early drop in susceptibility below 580°C is attributed to the successive appearance of grain size-specific T_Cs. Indeed, T_Cs between 500 and 560°C have been described by Haigh (1957) and Kobayashi (1959) for fine particles of magnetite. Both authors studied the remanence acquisition process of ultrafine magnetite crystallites during growth.

According to this interpretation, clusters containing larger grain size fractions should show a T_V and a sharp drop at T_C = 580°C. In fact, this is exactly what is observed in many χ-T curves (as shown for example in Fig. 2.2.3f). The associated FORC diagrams (Fig. 2.2.5b) confirm the dominance of a PSD fraction due to the compressed and non-symmetric contour, although SP grains are also present. Similar

FORC contours were published by Elbra et al. (2009) for the Eyreville suevites, but these authors did not give an explanation for the unusual mixing assemblage. Magnetite in these samples apparently contains an adequate fraction of crystals with multiple magnetic domains. Hence, variations of the magnetic behaviour within secondary magnetite associations are mainly controlled by the maximum grain size occurring in a particular cluster. If the maximum size is mainly below the SD threshold size of magnetite (0.1 µm), T_V is missing and T_C appears between 500 and 560°C. In the following, such clusters will be referred to as "subtype I" (mt sec I). If a sufficient fraction of larger, MD grains is present, T_V emerges and T_C appears at 580°C. These magnetites are named "subtype II" (mt sec II). Both types reflect a specific time interval within which goethite, the needle-shaped precursor, was able to grow. Assuming a constant growth rate, this time interval was shorter for subtype I than for subtype II.

In many curves, the susceptibility drop exceeds the typical T_C of stoichiometric magnetite (e.g. Fig. 2.2.3e) and indicates the presence of maghemite (e.g. Özdemir and Banjaree, 1984) or larger amounts of impurities (Michel et al., 1951). As it has been discussed for shocked magnetite, non-stoichiometry is an important feature which can lower or even suppress T_V. Indeed, LT_Vs and broadened transitions, which appear over the temperature interval between LT_V and T_V (Fig. 2.2.10), are visible in many curves. Within a given cluster, even the MD grains have a rather small grain size and each grain has only little contact with its neighbour. The surface/volume ratio is therefore quite high and makes the grains susceptible to oxidation, at least at the surface. However, no evidence for alteration is found from optical inspection in the SEM photographs, indicating that this process did not pervasively affect the magnetite clusters. Besides, the SAED pattern recorded by TEM on a magnetite monocrystal lack of supplemental reflections indicating oxidized magnetite or maghemite (Fig. 2.2.3d). However, it is known from various experiments that oxygen from air is sufficient to form a thin maghemite layer on the surface of magnetite grains (Aragón et al., 1985; Özdemir et al., 1993; Pardoe et al., 2001). Such a layer is hard to detect by microscopic inspection only. Especially for small particles, it can make up a distinct percentage of the bulk volume and therefore affect the magnetic properties. If this happened to secondary magnetite from the CBIS, why are these grains lacking of any additional reflexes in the SAED pattern? A likely explanation may be related to the preparation method of the FIB lamellae from which the SAED pattern was obtained. Since the maghemite layer is suggested to be restricted to the grain surface, it will be removed during preparation on the top and bottom side of the lamellae by the iron beam. In order to obtain solely reflexes from the monocrystal, the TEM electron beam is focussed onto the core of the grain during analysis and will thus exclusively transmit unoxidized material. Experiments from Aragón et al. (1985) and Özdemir et al. (1993) confirm that the Verwey transition is lowered or depressed for surface-oxidized magnetites. This effect is strongest for small grains and can completely depress the transition or, as seen for shocked magnetite, produce a LT_V. Depression, broadening and lowering of T_V in secondary magnetite can therefore be attributed to surface oxidation and the variation of maximum grain sizes between the magnetite clusters in the suevite sediments.

2.2.4 Conclusions

Three different generations of magnetite occur in the lithological units of the CBIS. Each generation is characterized by unique magnetic properties. Primary magnetite occurs in all granites and the basaltic dyke and it typically shows low a M_{rs}/M_s ratio and a sharp drop at T_V in the LT remanence curves. All rocks containing primary magnetite have not been significantly modified by shock and the general rock magnetic properties reflect the one of the pre-shock state. Shocked magnetite, in contrast, has smaller, but strongly varying grain sizes and shows a more PSD-like behaviour. Differences in the magnetic behaviour compared to primary magnetite mostly arise from grain size reduction and alteration. It is difficult to precisely separate the effects of shock deformation from those of oxidation by means of the particular magnetic parameters, since both processes trigger similar changes of the magnetic properties. Furthermore, post-shock alteration additionally modifies the shock deformation features. This observation is particularly important for natural impact craters, since those are usually affected by large post-shock hydrothermal events. In these settings, shock effects are most likely modified or erased to some extent. Since magnetic properties are very sensitive to these modifications an unequivocal interpretation of the rock magnetic properties in natural impact craters is very difficult and requires a combination of rock magnetic and microstructural investigations.

The presence of two Verwey transitions in shocked magnetite allows a division into three subtypes: (1) subtype I containing solely a LT_V, (2) subtype II containing solely a regular T_V and (3) mixed magnetite containing both T_Vs. T_V is mainly dependent on the particular oxidation state, which is again dependent on the surface/volume ratio. Since oxidation has a larger impact on smaller grains, T_V in these grains is more easily depressed and lowered. Subtype I is therefore characterized by a main fraction of grains in the PSD-size range, whereas subtype II contains mainly larger MD grains. Based on the assumption that the smaller grains in the shocked impact units are mainly a result of impact-related grain reduction, the subtypes are basically an indicator for shock-induced grain size reduction.

Secondary magnetite forms clusters of needle-shaped crystals ranging from SP to MD grain sizes. The crystal shape has been adapted from an iron oxide precursor, which was most likely goethite. Surface oxidation is evident from the magnetic measurements, but is not visible by optical inspection. The strong response of T_V in small grains to oxidation causes a lower transition temperature. Since reduction of T_V decreases with increasing particle size (Aragón et al., 1985), the additional presence of larger particles causes the transition to broaden. Considering the specific maximum grain sizes, the clusters can be subdivided into two subtypes: Subtype I mainly consists of SP and SD grains and shows no T_V as well as strongly reduced T_{CS}. The latter arise from nm-sized crystallites that are just large enough to enter the SD size range. Subtype II contains SP, SD and MD grains. The latter fraction is sufficiently large to induce PSD behaviour. These grains show a LT_V as well as a regular T_V, and sometimes the transition is broadened between LT_V and T_V. The strongly restricted grain sizes in subtype I indicate that the time interval during which the initial goethite growth proceeded was shorter than the one of subtype II. Despite their differences, both subtypes cluster on the SD-MD line of the Day plot and cannot be distinguished from each other in this diagram. The formation of magnetite clusters seems to be a typical result of hydrothermal mineralization since similar magnetite clusters have been found by Vahle et al. (2007) as secondary minerals in basalts.

An important result of this study is that the formation of LT_V in shocked and secondary magnetite can be attributed to the same mechanism, although the origin of both types is entirely different. Effects on T_V can be traced back to the presence of a non-stoichiometric volume fraction, which in both cases has been produced by oxidation. However, if oxidation is the main factor controlling these features, why does mixed magnetite form two distinct transitions at approximately fixed temperatures, whereas the transition in secondary magnetite is often broadened between LT_V and T_V? Here again, the particular surface/volume ratio may gives a reasonable explanation. MD grains in secondary magnetite can form considerable fractions but their maximum size is still restricted and does not exceed a few μm. In contrast, the average grain size of shocked magnetite is distinctly larger. As shown by Aragón et al. (1985) and Özdemir et al. (1993), the effect of oxidation on T_V decreases with increasing grain size and has therefore the strongest effect on small grains. In the grain size range of secondary magnetite, size variations of the small single grains are sufficient to slightly change the particular temperature at which T_V appears. Clusters showing a large variation of single grain sizes then develop a broad peak at T_V, indicating the progressive appearance of all T_Vs in this temperature range. The larger grains of shocked magnetite, however, are considerably less sensitive regarding their grain size. Indeed, the gneiss χ-T curve containing mixed magnetite (Fig. 2.2.1d) shows that both peaks are clearly separated from each other. Alteration has lowered T_V of grains below a certain threshold grain size, whereas T_V of larger grains remains unaffected. As described above, mixed magnetite in the suevite is affected by stronger alteration and therefore the response of T_V is again intensified, which is visible by a slight broadening of both peaks. The crucial factor controlling the response of T_V to oxidation is therefore the ratio between the oxidized and stoichiometric volume fraction.

Surprisingly, the FORC diagrams of secondary magnetite indicate that magnetic interactions within the clusters are negligible. Clustered associations of magnetite are usually known to strongly interact (e.g. Chen et al., 2007, Goya et al., 2003). In biogenic particles, the main reason for interactions is the chain-like particle arrangement with remanence vector alignments being in the same direction (Moskowitz et al., 1988). The random particle orientations in the samples of this study seem to strongly lower these interactions. Another effect may be that SD grains, which are suggested to produce the strongest interactions, are separated from each other by SP and PSD grains. The resulting spatial distance between the SD grains may additionally lower particle interactions. Nevertheless, the fraction of SD grains seems to be rather low, which is additionally indicated by the fact that a stable NRM direction is rarely present in samples containing secondary magnetite (see chapter 2.1). Random orientations of SD grains controlled by shape anisotropy should not give reproducible directions since the remanence vector in these particles is aligned along the long axis (Jackson et al., 1998).

The complicated interactions between various structural and magnetic features highlighted in this study reveals that it is important to combine magnetic and microstructural analyses. For example, secondary and shocked magnetite are difficult to distinguish merely by means of their magnetic properties since both generations have common magnetic features. In contrast, visual inspection allows one to clearly separate both types. The particular magnetic properties can then be interpreted on the basis of their specific appearance, which includes shape, alteration stage and microstructures. The fact that two different and locally separated generations of magnetite appear in the same unit shows that interpretations concluded only by magnetic measurements ought to be treated very carefully. In the CBIS, shocked and secondary magnetite have both lowered T_Vs, similar hysteresis parameters and even the LT behaviour of both is not easy to distinguish. Even if oxidation of a small grain size fraction is the common process affecting both types, each one has an entirely different geological background. Whereas small grain sizes in secondary magnetite are growth-related, the small grain size fraction in shocked magnetite is mainly a product of shock. Taking furthermore into account that both types occur within one single unit, e.g. the suevite, it becomes clear that magnetic measurements alone cannot resolve the complexity of such settings. The difficulties in interpreting the obtained data show that the application of various magnetic and microstructural methods need to be performed and combined, as it has been done in this study. This conclusion is especially important for natural environments, which contain a large spectrum of different rocks and magnetic sources, as it is the case for example in impact craters.

2.3 Iron-deficient pyrrhotite in the suevite from the CBIS

Abstract

In this chapter, shocked pyrrhotite from the suevite of the Eyreville drill core is examined with respect to its microstructures and magnetic behavior. The specific properties are compared to those of unshocked pyrrhotite, which mainly occurs in the schist block. Pyrrhotite in the schist block has been brittle deformed during the impact, but its magnetic properties remain rather unaffected from this process. Pyrrhotite from the suevite, in contrast, holds large amounts of stacking faults with average fault distances of 10 nm and deformation twins parallel to the hexagonal [001] direction. This pyrrhotite is strongly iron-deficient with an average metal/sulfur ratio of 0.81, indicating a higher vacancy concentration compared to the regular 4C structure. The chemical composition is far below the documented stability of the pyrrhotite group and coincides best with the one of smythite (Fe_9S_{11}). Iron-deficient pyrrhotite shows a 34 K transition, which is either strongly depressed and broadened, or totally absent. Its Curie temperature lies between 350 - 365°C. The structure of this mineral is based on a NiAs subcell, which is typical for all pyrrhotite modifications, but large streaking in the SAED pattern does not allow determination of the superstructure. Although large quantities of defects are present, an underlying ordered structure must be present and give rise to the ferrimagnetic behavior. It is suggested that ferrimagnetism arises from faultless domains of about 10 nm in diameter, which is probably at the lower limit of the single domain (SD) size and near to the threshold below which superparamagnetic behavior occurs. Since these domains contain an atomic ordering different to that of 4C pyrrhotite, some kind of vacancy ordering must have occurred during or after the transformation from pyrrhotite to iron-deficient pyrrhotite. Homogenous composition of iron-deficient pyrrhotite indicates that this transformation has developed pervasively. This observation makes it hard to conclude that the iron-deficient structure established after shock, since the high density of stacking faults is in conflict with such an interpretation. A more likely interpretation arising from the results of this section is that pyrrhotite transformation occurred already in the pre-shock rocks. The lattice defects have then been introduced into the iron-deficient phase and superimposed onto the oxidized smythite structure.

2.3.1 Introduction

From the magnetic carriers occurring in impact craters on Earth, magnetite (e.g. Vredefort: Henkel and Reimold, 2002; Morokweng: Henkel et al., 2002; Ries: El Goresy, 1964) and monoclinic pyrrhotite (e.g. Bosumtwi: Kontny et al., 2007; Sudbury: Vaughan et al., 1971) are the most important ones. Pyrrhotite is also assumed to play an important role for the magnetization of meteorites (Garrick-Bethell and Weiss, 2010; Rochette et al. 2003, 2005; Stöffler et al., 1986) and the Martian crust (Kletetschka et al., 2000; Louzada et al., 2007, 2011; Rochette et al., 2003). A ferrimagnetic to paramagnetic transition of Fe_7S_8

has been reported to take place at about 2.8 GPa (Rochette et al., 2003), and at~6.8 GPa (Kamimura et al., 1992; Kobayashi et al., 1997) under hydrostatic pressure. It has been suggested that the transition is linked with the Hugoniot elastic limit of pyrrhotite occuring around 3.5 GPa (Louzada et al., 2007, 2010). The material reacts by elastic deformation below that pressure whereas plastic deformation takes over above. Among the basic features measured after shock, an increase of the low temperature memory, bulk coercivity and saturation isothermal remanent magnetization (SIRM) are the most prominent ones (Louzada et al., 2007). These changes are mainly attributed to the fragmentation of large multidomain (MD) grains into smaller single domain (SD) grains (Gilder et al., 2004; Louzada et al., 2005) due to mechanical brecciation (Williamson et al., 1986). Louzada et al. (2007) observed in shock experiments on pyrrhotite that the loss of magnetization induced by shock reached up to 90%, meaning that 10% of the initial magnetization remained in the shocked minerals.

In natural environments, alteration can strongly modify pyrrhotite and initiate transformation to pyrite, marcasite (both: FeS_2), smythite (Fe_9S_{11}) or even magnetite (Fe_3O_4), hematite (Fe_2O_3) and goethite (FeOOH) (Dekkers, 1990; Fleet, 1978; Janzen et al., 2000; Jover et al., 1989; Krs et al., 1992; Mang et al., 2012). Oxides are favored by the production of Fe^{2+} to Fe^{3+}, since the latter is preferably bonded to oxygen rather than to sulfur in such reactions (Pratt et al., 1994). Oxidation generally starts at the grain surface and progressively triggers diffusion of Fe^{2+} from inner to outer grain parts. However, this process will break up if the grain is sufficiently large and a maximum thickness of the oxidized rim is reached (Janzen et al., 2000). Since all listed secondary phases, except smythite, are not ferrimagnetic, alteration tends to generally diminish the magnetic behavior of pyrrhotite assemblages. In the case that smythite is formed, the initial remanence is modified into a CRM.

2.3.2 Results

Optical and electron microscopy

Pyrrhotite is exclusively present in the Eyreville core and occurs therein in the schist block (Fig. 2.3.1a) and the suevite. The schist block contains large MD pyrrhotite grains consisting of hexagonal (NC) and monoclinic (4C) modifications. The grains are strongly fractured and altered (Fig. 2.3.1b). Previous ductile deformation is clearly visible. As illustrated in chapter 2.1, brittle deformation is related to the impact event, but no higher shock stages are documented for this unit. This unit will be therefore used as reference material referring to as "unshocked pyrrhotite" in this section in order to better understand the magnetic behavior of shocked pyrrhotite.

Shocked pyrrhotite usually occurs as weathered single grains and rarely within schist fragments in the Eyreville suevite (Fig. 2.3.1c). Single grains are often associated in clusters, but the grains are mostly separated from each other. These clusters are likely the result of brecciation and subsequent alteration of the initial, larger grains (Fig. 2.3.1d). They consist of ferrimagnetic pyrrhotite, but the twinning structure, which is typical for pyrrhotite (Janzen et al., 2000; Kontny et al., 2000) is missing. The pyrrhotite grains

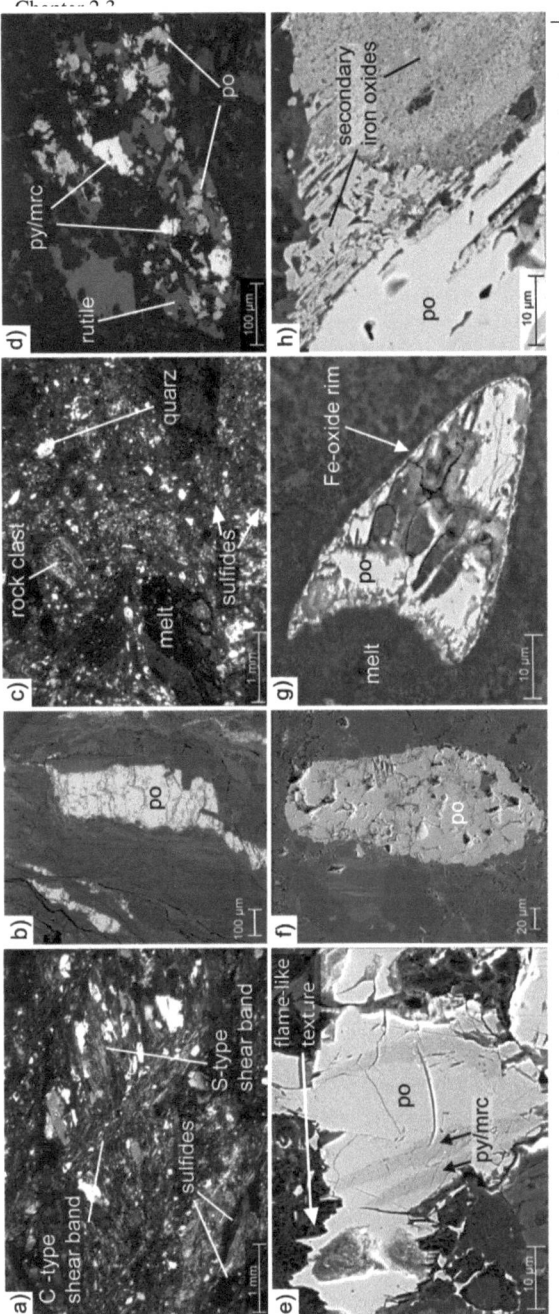

Figure 2.3.1: Pyrrhotite types in the Chesapeake Bay impact structure (CBIS). a: typical fabric of the schist rock pre-impact C- and S-type shear bands (Gibson et al., 2009). The sulfides including pyrrhotite are elongated in line with the general metamorphic fabric. OM image. b: Pyrrhotite grain in the schist. Besides ductile deformation, impact-related brittle deformation is clearly visible. BSE image. c: Typical fabric of the suevite. Fragments of different sizes occur within the matrix, which consists of fine rock particles. Fragments are remnants of rocks, but also crushed minerals and impact melt. Pyrrhotite occurs in this extremely unsorted sediment as single grains in the matrix or within melt fragments. OM image. d: Pyrrhotite (po: reddish-brown colour due to coating with ferrofluid) and pyrite/marcasite (py/mrc) grains are surrounded by iron oxide. The assemblage documents the alteration of pyrrhotite to pyrite/marcasite and finally to iron oxide. OM image. e: Pyrrhotite/FeS$_2$ intergrowth. The flame-like texture arises from the anisotropic crystal structure and is visible if an altered grain is cut approximately parallel to the c-axis. BSE image. f: Pyrrhotite from the sample CB 30 is less fractured and reveals larger grain sizes compared to pyrrhotite from other samples in the suevite. SE image. g: Pyrrhotite occurring within a melt fragment. Heat treatment is clearly visible and secondary oxidation has formed a rim of Fe-oxide surrounding this fragment. BSE image. h: Strongly oxidized pyrrhotite. To the left fractured pyrrhotite is present but this grain is transformed to iron oxide, which appears in the center. The iron oxide shows a skeletal structure, which results from the former pyrrhotite buildup. The skeletal structure reveals that oxidation preferably occurs along preferred lattice orientations of pyrrhotite. On the right hand side of the photograph another iron oxide with a different structure is present. It replaces both pyrrhotite and its oxidation product. BSE image. OM: optical microscope; BSE: back scattered electron; SE: secondary electron; po: pyrrhotite; py: pyrite; mrc: marcasite; FIB: focused iron beam (used for TEM preparation).

shows a strong optical anisotropy under reflected light. Nevertheless, lamellae structures are visible in larger grains and illustrate an intergrowth between ferrimagnetic pyrrhotite and pyrite, respectively marcasite (Fig. 2.3.1e). Pyrite and marcasite have formed during alteration and the lamellae structure shows that diffusion is favored along specific crystallographic directions. For the 4C structure, these directions lie perpendicular to the c-axis along which Fe-bearing layers are stacked (Putnis, 1992). Alteration of this anisotropic structure furthermore produces serrated grain rims, which run parallel to the c-axis. If the grain is cut along the same direction, a flame-like texture appears at the rim of these grains (Fig. 2.3.1e) One suevite sample, CB 30 (Fig. 2.3.1f), holds larger grains that seem to be less affected by alteration and fracturing. Some grains occur within melt fragments (Fig. 2.3.1g) and are partially molten. These grains are often surrounded by a rim of iron oxide, which developed during oxidation of the former grain rim. Transformation of pyrrhotite to secondary iron oxides is a general abundant feature and pervasively developed in some grains, whereas others seem to be rather intact. In the former case, pyrrhotite occurs as relict, surrounded by secondary iron oxide phases (Fig. 2.3.1h). Some such iron oxides show a skeletal form, which is a remnant of the former pyrrhotite structure.

TEM photographs of pyrrhotite reveal partial transformation from pyrrhotite to marcasite (Fig. 2.3.2a) along specific directions. According to SAED pattern, marcasite shows a preferred orientation relative to pyrrhotite with c_{po} ∥ a_{mrc} and a_{po} ∥ c_{mrc} (based on a hexagonal pyrrhotite lattice, Fig. 2.3.2a + b). The interface between the two phases is corrugated (Fig. 2.3.2a). These orientation relationships are consistent with observations by Fleet (1978) using X-ray diffraction. Small angular deviations from the ideal orientation relationship arise from slight misorientations among marcasite domains, forming a fibrous or slightly columnar texture. Internally the 30 to 100 nm-sized marcasite domains appear well crystallized with no signs of shock-induced defects. Most of the observed secondary sulfide oxidation products consist of marcasite, but pyrite can also be present. Pyrite forms as direct oxidation product of pyrrhotite or by inversion of marcasite (Murowchick, 1992). Single pyrite grains, which are abundant in the thin sections, are assumed to be of secondary, post impact origin as they overgrow the suevite matrix.

Looking at the crystal lattice of pyrrhotite, it appears that strong deformation is visible. TEM brightfield images in Fig. 2.3.2c and 2.3.3 show various types of lattice defects including high densities of stacking faults. The density of stacking faults parallel to hexagonal (001) is extremely high with average fault distances in the order of 10 nm. These most probably shock-induced faults represent disturbances of the regular ABAB hcp stacking sequence of sulfur layers and are connected to a high density of partial dislocations. Superposed on this stacking fault microstructure, twinning can be observed (Fig. 2.3.2d). The widths of twin domains are on the order of several 100 nm and oriented such that their interfaces are subparallel to (001). Their domain boundaries are not clearly defined, do not show a constant width as it is the case for undeformed pyrrhotite (see e.g. Fig. 2a in Kontny et al., 2007) and are not parallel to each other such as in undeformed pyrrhotite. The domains in shock-deformed pyrrhotite probably arise from partial rotation of the vacancy-bearing iron sublattice similar to that described for the twinned 4C modification (Putnis, 1975) and are of a mechanical origin.

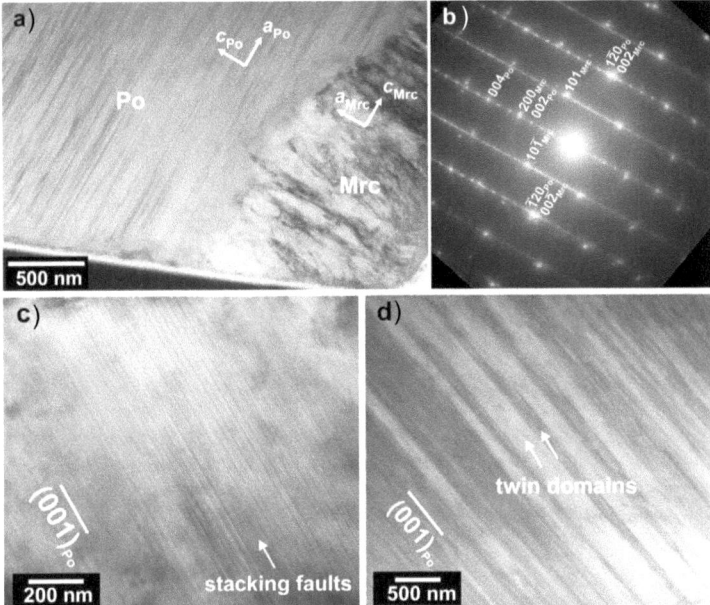

Figure 2.3.2: TEM photographs of shocked pyrrhotite. a: Bright field TEM image showing an epitactic growth of marcasite on pyrrhotite. Pyrrhotite shows abundant stacking faults and twin domain boundaries and marcasite displays a fibrous texture due to slight misorientations between crystal domains. Note the diffuse deformation twins in pyrrhotite. Corresponding axes of both minerals are marked. b: diffraction pattern of the grain contact in a. Crystal lattices of pyrrhotite and marcasite share a common orientation with slight angular deviations of the marcasite lattices due to subgrain boundaries. c: Stacking faults occur along (001). The dark-light contrast indicates lattice defects. d: Twin domains along (001). The domains cannot be clearly distinguished from each other and contain further smaller subdomains. The planes are tilted slightly off the zone axis. Likely the presence of stacking faults influences the Fe vacancy arrangement and causes the blurred boundaries between twin domains.

The SAED patterns in Fig. 2.3.4 reveal a partial consistence with a 4C diffraction pattern. Within the row of fundamental NiAs reflections, the SAED pattern strongly resembles the one of 4C-pyrrhotite. However, reflection doublets within the intermediate row do not fit any low temperature 4C or NC pyrrhotite phase (e.g., Dódony and Pósfai, 1990; Harries et al., 2011; Pósfai et al., 2000). Indexing of superstructure reflections is strongly hampered due to the streakiness induced by strong disorder. The intermediate row in the SAED pattern represents a unique superstructure that is probably arising from the partial disorder of Fe vacancies. If reordering occurred, this process did not proceed to completeness. The presence of abundant stacking faults excludes any advanced annealing processes and the measured c-lattice constant of 5.67 Å for the basis reflections (outer rows in Fig. 2.3.4a, corrected using marcasite reflections as internal standard) is consistent with the NiAs elementary cell. Diffuse streaking of the reflections forming the inner row makes it impossible to determine the superstructure multiplicity. As

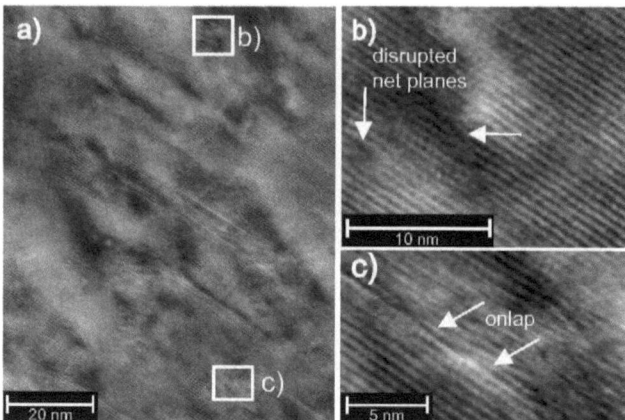

Figure 2.3.3: Bright field TEM micrographs of shocked pyrrhotite. a: View along a hexagonal [210] equivalent zone axis (Harries et al., 2011). Stacking faults appear as dense and highly planar features with lateral spacing of only a few nm. These planes are interpreted to represent displacements within the sulfur sublattice. b + c: magnification of a). Various crystal defects are visible along the lattice planes.

the NiAs cell parameters are the basis parameters of every pyrrhotite type, the presented measurements confirm the presence of pyrrhotite but leave open the superstructure type.

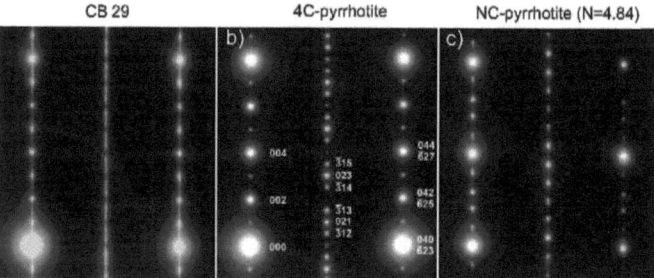

Figure 2.3.4: SAED pattern of pyrrhotite seen along a hexagonal [210] equivalent zone axis. a: pyrrhotite of the Chesapeake Bay suevite. The outer rows matches with 4C pyrrhotite. The reflections of the inner row deviate from the 4C modification and reflect a superstructure. The blurred reflections derive from partial vacancy disorder. b: Reference pattern of stoichiometric twinned 4C pyrrhotite. c: Reference pattern of NC pyrrhotite. The visible reflections do not match with pyrrhotite from the suevite in the CBIS. SAED: Selected area electron diffraction.

Electron microprobe analysis

Electron microprobe analyses of pyrrhotite from four suevite samples and one reference schist sample are presented in Table 2.3.1. Fig. 2.3.5 shows discrimination diagrams of the obtained data. These

Chapter 2.3   Iron-deficient pyrrhotite

Table 2.3.1: Average values of microprobe analyses of shocked and unshocked pyrrhotite. Co content is below 0.15 at.%.

sample	depth [m]	n	Fe [at.%]	st. dev.	S [at.%]	st. dev.	Ni [at.%]	st. dev.	T_C [°C]	Fe + Co+ Ni [at.%]	Fe/S ratio	st. dev.	(Fe + Ni + Co)/S ratio	st. dev.
schist														
CB 32	1562.6	123	45.99	(259)	53.579	(241)	0.13	(021)	320	46.221	0.858	(008)	0.863	(008)
suevite														
CB 25	1404.5	87	44.417	(229)	55.064	(237)	0.170	(045)	364	44.667	0.807	(006)	0.811	(067)
CB 29 1b	1421.9	11	43.985	(221)	55.073	(486)	0.481	(062)	357	44.547	0.799	(010)	0.809	(011)
CB 29	1421.5	70	43.673	(231)	55.153	(239)	0.806	(034)	349	44.553	0.792	(007)	0.808	(007)
CB 30	1442.4	36	46.024	(307)	53.313	(261)	0.109	(016)	325	46.302	0.863	(009)	0.869	(009)

n: number of measured points, at.%: atomic percent; st. dev.: standard deviation; T_C: Curie temperature

diagrams help to recognize that pyrrhotite of all studied samples, including the reference sample, shows iron contents that are below the compositional range of 4C pyrrhotite (46.4 - 46.9 atomic %; Fe_7S_8: 46.6 atomic % Fe; Kissin et al., 1982). Pyrrhotite from the suevite is strongly depleted in iron (44.55 to

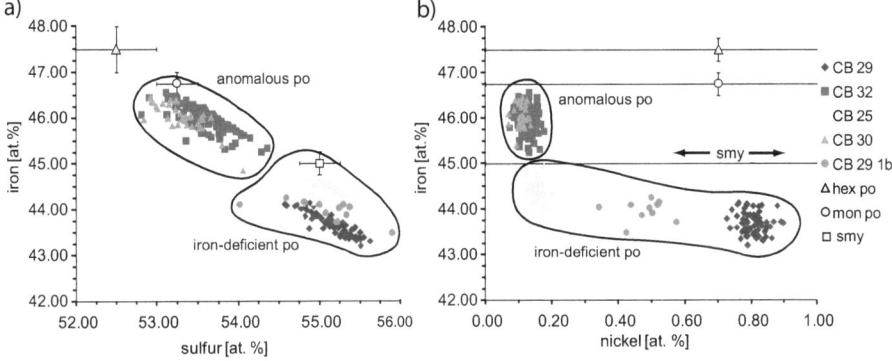

Figure 2.3.5: Discrimination diagrams of iron vs. sulfur (a) and iron vs. nickel (b) concentration of pyrrhotite. Two distinct groups can be discriminated: one is related to iron-deficient and one to anomalous pyrrhotite, respectively. Iron deficient pyrrhotite displays a distinctly higher variability in nickel concentration. Reference data are shown for smythite (smy), monoclinic pyrrhotite (mon po: Fe_7S_8) and hexagonal pyrrhotite (hex po: Fe_9S_{10}). Bars indicate the compositional range of the particular element.

44.66 atomic %). Even if the sum of Fe + Ni + Co is considered, the metal concentration (between 44.55 and 44.67 atomic %) of this pyrrhotite is too low for any known pyrrhotite modification. One exception, however, is the suevite sample CB 30, which reveals iron concentrations similar to the ones from the schist (sample CB32). Sulfide minerals in this section are less altered compared to those from the other sections. In general, iron concentrations slightly vary within one section, but are constant within a single grain or grain cluster. Fig. 2.3.6 shows a typical iron distribution within an iron depleted pyrrhotite grain. All rims show thin zones of further iron depletion induced by oxidation, but within the inner part, the iron concentration is homogeneous. Besides the thin grain rims, no indications of chemical zonation or

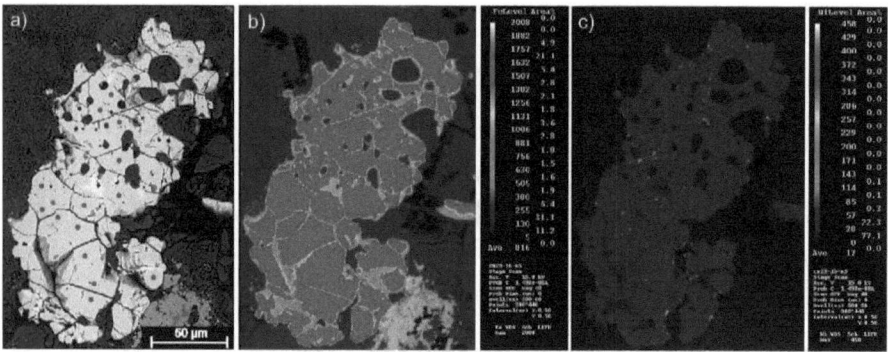

Figure 2.3.6: Element distribution map of an iron-deficient pyrrhotite grain (sample CB29) showing significant depletion of iron (see Fig. 2.3.5). a: SEM photograph of the grain. Red dots mark the points analyzed by electron microprobe. b: Iron distribution map. No indications of diffusion or zonation are visible in the inner grain parts. c: Nickel distribution map. Local Ni-accumulations may indicate tiny Ni-bearing sulfides.

diffusion processes were observed within the inner grain parts (Fig. 2.3.6b), wherefore a thin intergrowth of FeS_2 and pyrrhotite is unlikely. Highly variable Ni concentrations are characteristic throughout all measured grains (Table 2.3.1, Fig. 2.3.6). The tiny spots with higher Ni contents appearing in Fig. 2.3.6c probably indicate local exsolution of Ni-bearing sulfides like pentlandite.

Magnetic behaviour

Temperature-dependent magnetic susceptibility measurements (χ-T curves) show several magnetic transition temperatures (Fig. 2.3.7). The schist shows a distinct increase of susceptibility at about 220°C (Fig.2.3.7a), which represents the transition from antiferromagnetic NC to ferrimagnetic NA pyrrhotite (λ-transition; Schwarz and Vaughan, 1972). At 325°C, a sharp drop marks the Curie point (T_C) of ferrimagnetic 4C pyrrhotite. Similar χ-T curves are reported in Kontny et al. (2000) for mixed hexagonal and monoclinic pyrrhotite modifications with a dominating contribution of the monoclinic type. Pyrrhotite from the suevite, in contrast, shows no increase of susceptibility before reaching the Curie point (Fig. 2.3.7b). This behavior confirms the absence of hexagonal pyrrhotite. However, the T_C of all iron deficient pyrrhotite grains is shifted towards higher temperatures. These T_Cs lie between 350 and 365°C (Fig. 2.3.7). From the five investigated pyrrhotite-bearing suevite samples only sample CB 30 (remembering that pyrrhotite in this sample contains a similar composition than the reference sample), shows a T_C of about 325°C, typical for the ferrimagnetic 4C pyrrhotite. In general, pyrrhotite from all samples including the reference sample are not stable during temperature treatment up to 700°C and transform at~500°C into a secondary magnetite or maghemite phase with T_C between 520 and~635°C. This is visible in the cooling curve, where the absence of the pyrrhotite T_C marks the instability of this phase during

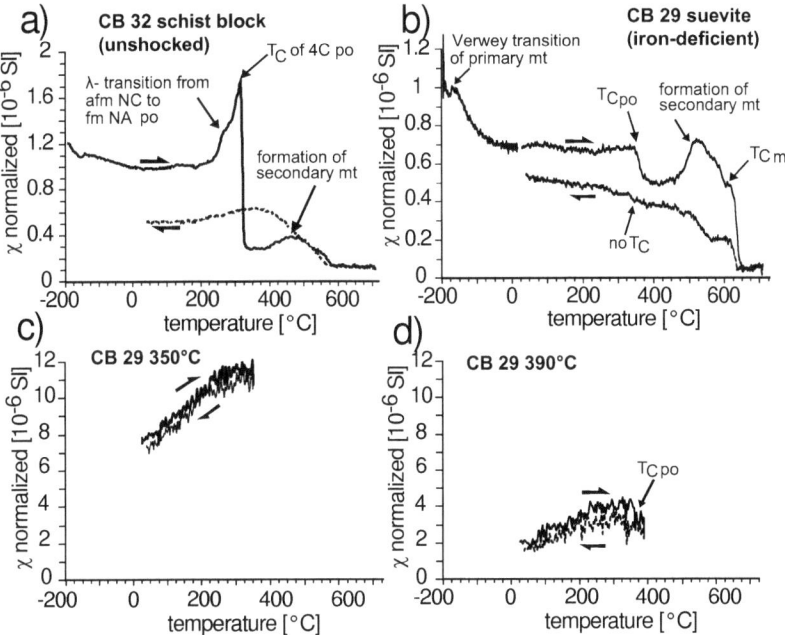

Figure 2.3.7: Normalized susceptibility as a function of temperature between -200°C and 700°C. Heating and cooling curves are marked by arrows. a: pyrrhotite from the unshocked schist. Transitions at 220°C and 325°C reveal the presence of hexagonal and monoclinic pyrrhotite. Irreversible behavior of heating and cooling curve is attributed to the formation of secondary magnetite during heating. Magnetic susceptibility (χ) used for normalization at 27°C = $14 \cdot 10^{-3}$ SI unit. b: The suevite contains magnetite (mt) and iron deficient pyrrhotite. T_C is distinctly higher than in a) and lies now at ~360°C. Most of the iron depleted pyrrhotite is oxidized to an iron oxide and is thus no longer visible in the cooling curve. χ used for normalization at 27°C = $13.4 \cdot 10^{-3}$ SI unit. c: Heating of iron-depleted pyrrhotite up to 350°C and back to room temperature. No T_C is visible. d: Same experiment as c) using another sample and heating up to 390°C. The susceptibility drops irreversibly at~365°C indicating T_C. afm: antiferromagnetic; fm: ferrimagnetic.

heating even though measurements were conducted in argon atmosphere.

In order to learn more about the stability of pyrrhotite in the suevite, χ-T curves with stepwise heating up to different temperatures (300, 320, 340, 360, 380 and 700°C) and subsequent cooling where performed. Two different experimental designs were applied: (1) the same sample was used for all runs, and (2) each run was made with a fresh sample. In the first series of experiments, independent of whether schist or suevite samples were used, no clear pyrrhotite-related T_C was detected in any run. The last run (700°C) only showed the presence of a maghemite-near phase, which was probably again the product of unstable pyrrhotite behavior. During the second experimental design, the susceptibility dropped noticeably around 360°C (Fig. 2.3.7c and d). This decrease, indicating T_C of the iron sulfide phase, was

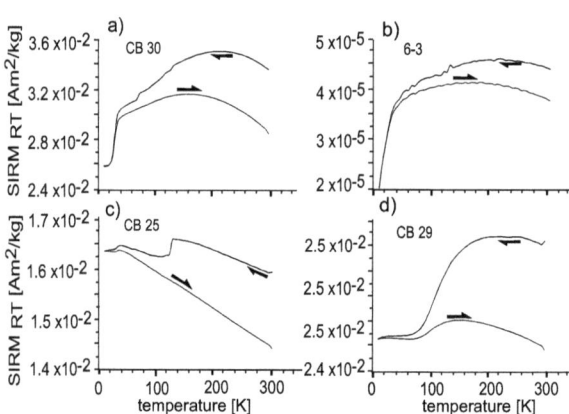

Figure 2.3.8: Low-temperature cycling of saturation isothermal remanent magnetization (SIRM) given at room temperature. a: Suevite with monoclinic 4C pyrrhotite. A transition at 34 K is visible by a distinct remanence increase. b - c: Suevite containing iron-deficient pyrrhotite. b: Although the remanence increase is still visible it is much smaller (note the different scale in y-axis compared to (a)), broadened and reversible. c: The transition is completely absent. The remanence increase at~100 K can be related to a magnetite-near phase. d: The transition is present only as a small increase in remanence and only half of the remanence is measured at 34 K on the heating curve. The transition occurs over a larger temperature interval.

present in all runs with maximum temperatures above 360°C. This experiment therefore shows that all pyrrhotites are metastable phases, which are oxidized if repeatedly heated above~300°C. T_C exclusively appears on the first heating.

During low temperature (LT) cycling, sample CB 30, whose stoichiometry is comparable to the reference sample, shows a distinct remanence drop at 34 K as a consequence of the 34 K transition. In iron-deficient pyrrhotite, this transition is not, or only poorly, developed. Fig 2.3.8b - d show the LT remanence curves of such samples. The transition is present, but shows a strong depression and broadening in sample 6-3 and CB 25 similar to what is observed at the Verwey transition (T_V) for oxidized magnetite (e.g. Özdemir et al., 1993). Remanence above the transition is distinctly less temperature-dependent in 6-3 than in CB 30. In CB 29, no transition that can be assigned to pyrrhotite is present. Anyway, CB 25 and CB 29 contain additional magnetite, which can be recognized by the remanence drop at T_V around 120 K.

2.3.3 Discussion

Anomalous behavior of "pyrrhotite" from the suevite

The formula of the pyrrhotite group is $Fe_{1-x}S$ with a compositional range between $0 < x < 0.125$ (e.g. Kissin and Scott, 1982). The most iron-deficient end member of this solid solution is thus $Fe_{0.875}S$ or Fe_7S_8, corresponding to an iron content between 46.4 and 46.9 atomic % Fe. The low metal concentration (46.22 atomic %) of pyrrhotite from the schist sample (Table 2.3.1) indicates the presence of "anomalous pyrrhotite" as defined by Clark (1966), which is characterized by its weaker ferrimagnetism compared to 4C pyrrhotite and its widespread occurrence as a secondary mineral. Magnetic measurements (Fig. 2.3.7a), clearly confirm the presence of antiferro- and ferrimagnetic modifications in the schist, which can be recognized by the particular transition temperatures (Schwarz and Vaughan, 1972). According to the present knowledge, this schist is the only pyrrhotite-bearing lithology of the CBIS and therefore the most likely candidate, which comes into consideration in terms of source and origin of shocked pyrrhotite in the suevite. Basaltic dykes are described from the Appalachian Orogen (Gibson et al., 2009; Horton et al., 1989) and may be a further source rock. However, only one basaltic dyke was drilled from the CBIS (chapter 2.1), but this unit exclusively contains magnetite.

Iron-deficient pyrrhotite from the CBIS suevite is characterized by properties that are in agreement with stoichiometric 4C pyrrhotite. Such congruencies are the strong anisotropy in reflected light, a bronze–red colour, ferrimagnetism at room temperature and a NiAs subcell (Fig. 2.3.4a). However, this "pyrrhotite" shows a significant iron deficiency far below the lower stability limit of the 4C structure and also far too low to agree with "anomalous pyrrhotite". Pyrrhotite grains from the suevite show metal/sulfur ratios between 0.808 and 0.811 (Table 2.3.1), yielding a mineral formula between $(Fe,Ni)_4S_5$ (0.80) and $(Fe,Ni)_9S_{11}$ (0.818). This composition is far away from the end member of the lower limit of the pyrrhotite system (0.875). Iron-deficiency is always linked to abnormal high T_Cs between 350 and 365°C. In an equaly manner, the 34 K transition is broadened and depressed or even absent. CB 30, however, contains grains with a composition close to pyrrhotite in the reference sample and a regular T_C. Alteration is distinctly less in these samples, as observed in the microscopic studies.

Formation of a smythite-like phase

The composition of iron-depleted pyrrhotite from the suevite coincides best with the composition of smythite $(Fe,Ni)_9S_{11}$ (Fig. 2.3.9). The crystal structure of smythite (trigonal) is also based on the hexagonal NiAs subcell and is therefore related to the broad range of pyrrhotite modifications (e.g. Vandenberghe et al., 1991). Furthermore, pyrrhotite and smythite show similar physical properties and the question remains open whether smythite should be defined as a separate phase in the $Fe_{1-x}S$ system or not. X-ray diffraction precession photographs of Fleet (1982) have shown many extra reflections for pyrrhotite-smythite intergrowth, which have been indexed on the base of a twinned smythite lattice. This twin plane has the same orientation as twin planes in 4C pyrrhotite (twin plane [001]). Broadened reflections forming diffraction streaks in lattice rows parallel to the c-axis of smythite have been equally described by Fleet (1982). The author ascribed this feature to stacking disorder in closely packed layers of atoms. In pyrrhotite grains from the CBIS suevite a high density of stacking faults and twinning parallel to the hexagonal c-axis occur with average fault distances in the order of 10 nm. The strongly shocked grains of quartz and feldspar occurring in the suevite (Horton et al., 2009a; Wittmann et al., 2009a) suggest a shock-induced origin of the defects in iron-deficient pyrrhotite. Its c-lattice constant is in agreement with the NiAs subcell, but complete reordering of vacancies seems to be inhibited by the numerous lattice

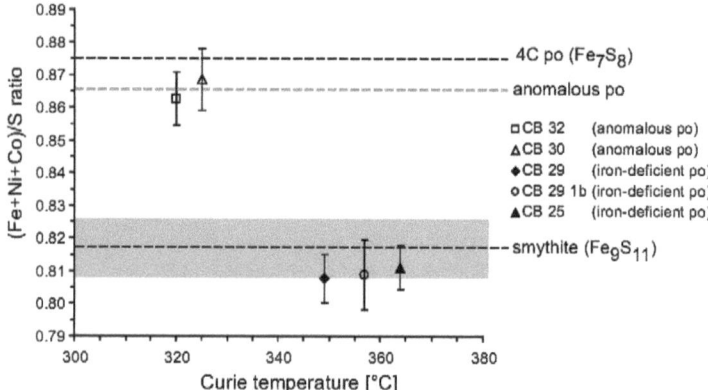

Figure 2.3.9: Average metal/sulfur ratio vs. Curie temperature. Both, (Fe,Ni,Co)/S ratio and Curie temperature differ distinctly between anomalous and iron depleted pyrrhotite. A reference line marks the composition of the respective mineral listed on the right hand side. The grey bar marks the compositional range of smythite after Taylor and Williams (1972).

defects and makes it impossible to determine a pyrrhotite superstructure. Nevertheless, ferrimagnetism of these grains at room temperature is in agreement with only some kind of vacancy ordered structure, since ferrimagnetism is impossible to arise from unstructured atomic arrangements (e.g. Bertraut, 1953). The total absence of a Curie point at 325°C excludes a partial conservation of unshocked 4C pyrrhotite in the studied iron-depleted pyrrhotite grains. Therefore the observed faultless domains in the range of about 10 nm must show a certain atomic ordering. Strong reduction or absence of the 34 K transition and T_Cs between 350 and 365°C indicate that this ordering is different to 4C pyrrhotite. Previous LT measurements on a smythite concentrate from Miocene sediments (Rochette et al., 1994) proved that no transition occurs at 34 K for this mineral (Fig. 2.3.10). Compiling all observations on iron-deficient pyrrhotite, it can be concluded that the iron-deficient pyrrhotite phase is similar to smythite. The smythite structure can be considered as a derivative of the NiAs subcell by introduction of regular stacking faults within the S sublattice and partial removal of Fe atoms (Fleet, 1982), which is in agreement with the observations of this study. A brief comparison between regular 4C pyrrhotite, smythite and iron-deficient pyrrhotite of this study is given in Table 2.3.2.

Smythite is described in the literature as a product of rapid cooling (Fleet, 1982; Furukawa et al., 1996) or low temperature oxidation (Taylor and Williams, 1972; Watmuff 1974). Rapid cooling of S-rich hexagonal 1C pyrrhotite from above 500°C forms smythite because this iron-deficient mineral is able to balance the low Fe/S ratio in the system (Furukawa et al., 1996). This process does not allow a co-existence of smythite with pyrite and pyrrhotite, which are the stable minerals at slower cooling rates. Iron-deficient pyrrhotite from the CBIS suevite is intergrown with marcasite and pyrite, but their occurrence in the Chesapeake suevite is obviously a product of secondary alteration (Fig. 2.3.1d + e and 2.3.2a)

Table 2.3.2: Comparison between magnetic and mineralogic features of 4C pyrrhotite, smythite and shocked pyrrhotite

	4C pyrrhotite	smythite	iron-deficient pyrrhotite
Formula	Fe_7S_8 ($Fe_{0.875}S$)	Fe_9S_{11} ($Fe_{0.818}S$)	$Fe_{0.808-0.812}S$
Magnetic properties	Ferrimagnetic Curie-T: 325°C[1] 34 K transition	Ferrimagnetic Decomposition-T:~380°C[2] No 34 K transition[3]	Ferrimagnetic Curie/decomposition-T: ~350 - 356°C No or strongly suppressed 34 K transition
Formation	(a) Crystallization in magmatic and (b) metamorphic rocks (c) hydrothermal formation	(a) Crystallization during quenching of hexagonal 1C pyrrhotite (b) Formation at the presence of siderite in sedimentary rocks (c) Low temperature oxidation	Shock-induced transformation of anomalous 4C pyrrhotite?
Lattice parameter	a = 6.865 Å, c = 22.72 Å	a = 3.465 Å, c = 34.34 Å	c = 5.67 Å
Crystal structure	monoclinic[a]	Hexagonal/trigonal	?
Textural/ mineralogical features	Often lamellae of hexagonal 1C pyrrhotite	Crystallization under quenching conditions occurs under absence of pyrite[4] flamelike grains in pyrrhotite (LT oxidation)[5]	No chemical zonation, no presence of pyrrhotite/marcasite are common
Ni-content	0 - 1.4 at.%	0 - 5.5 at.%	0.15 - 0.81 at.%

[1]Dunlop and Özdemir (1997) [2]Hoffmann (1993), [3]Furukuwa and Barnes (1996), [4]Taylor and Williams (1972), [a]all pyrrhotite superstructures are based on a hexagonal NiAs-type subcell with a = 3.45Å and c = 5.75Å.

rather than of cogenetic crystallization with pyrrhotite at high temperature. Most pyrrhotites of this study are strongly affected by brittle deformation and do not show any evidence of significant heating. The rare grains, having suffered partial melting, lack of any optical evidence indicating the formation of newly crystallized phases. These grains have not been further investigated by TEM.

LT oxidation of pyrrhotite is suggested to produce an elemental zonation within the affected grains. Pratt et al. (1994) presented an oxidation mechanism for pyrrhotite that involves the formation and diffusion of Fe^{3+} and results in an oxyhydroxide layer and a zone of strong iron depletion. Although the diffusion is at first powered by a chemical gradient, the supply of iron becomes weaker during progressive reaction, resulting in a final stagnancy. All studied pyrrhotite grains from the CBIS suevite lack any textural evidence indicating fluid-induced diffusion processes, which would have affected the whole grains (Fig. 2.3.6). Although secondary alteration to pyrite and marcasite is present, the unaltered parts of the grains show no variation in the Fe/S ratio and the standard deviations of pyrrhotite from the suevite are mostly smaller than the ones from the schist (Table 2.3.1). This observation makes it hard to explain the iron-deficiency exclusively by diffusion and oxidation processes, even if LT alteration below 100°C pervasively affected the suevite (Wittmann et al., 2009a). Furthermore, the high concentration of lattice defects is in conflict with such an interpretation. If only LT alteration formed the iron-deficient pyrrhotite, most of the deformation structures should have been annealed by reordering processes. This case is illustrated by the secondarily formed, well crystallized marcasite, as shown in Fig. 2.3.2a, b).

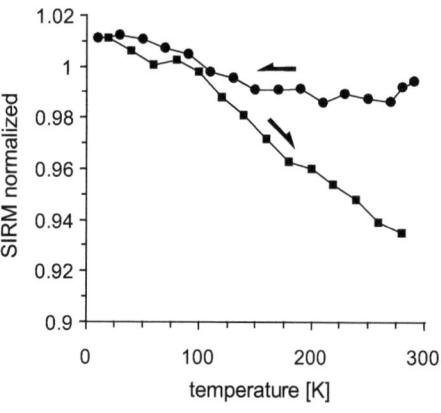

Figure 2.3.10: Low-temperature cycling of saturation isothermal remanent magnetization (SIRM) given at room temperature (3 T) of a smythite concentrate from Miocene sediments measured with a SHE cryogenic magnetometer in 20 K steps (data from Rochette et al. 1994). Magnetization is normalized to the initial value at room temperature. Magnetization increases continuously during cooling below 250 K, a 34 K transition is visible neither on the cooling, nor on the heating curve.

Regarding the Ni concentrations of pyrrhotite grains in the suevite it is observed that their typical concentration lies between 0.1 and 0.9 atomic %. This relatively large variation suggests that pyrrhotite was enriched of Ni either during or after shock. On one hand, higher Ni concentrations in pyrrhotite are in agreement with an origin from basaltic lithologies (Horton et al., 1989), which have been described for the basement of the CBIS. On the other hand, other grains with lower Ni concentrations coincide with those from the schist (Fig. 2.3.5b). Higher amounts of Ni in smythite (up to 5.5 atomic %) in comparison to pyrrhotite (0 - 1.4 atomic %) are described to suppress exsolution of Ni-rich phases like pentlandite in the Fe-Ni-S system (Taylor and Williams, 1972). Therefore, the tiny Ni accumulations (Fig. 2.3.6c) indicating exsolution of pentlandite seem to be of primary origin, in agreement with different lithological sources of pyrrhotite, and are presumably not related to shock-metamorphism and the formation of smythite.

Between 250°C and 340°C, smythite is known to decompose into pyrrhotite-like phases (Krs et al., 1992). χ-T curves of smythite measured by Hoffmann (1993) show an increase of susceptibility up to ~280°C followed by a decrease that ends at 380°C. Newly formed 4C pyrrhotite representing one of the decomposition products is then visible in the cooling curve (T_C: 325°C). The authors estimate the T_C of smythite to lie around 300°C. As Fig. 2.3.7b reveals, the χ-T curves from the here presented study look different. Susceptibility decreases around T_C of 350 - 365°C and no ferrimagnetic 4C pyrrhotite is visible in the cooling curve at 325°C. The non-reversibility of susceptibility as function of temperature indicates a strongly unstable phase, which does, however, not transform into 4C pyrrhotite but into ferrimagnetic iron oxides (presumably maghemite, see Fig. 2.3.7b).

Origin of iron-deficient pyrrhotite

Iron-deficiency and elevated T_Cs between 330 and 350°C have been equally observed in shocked pyrrhotites from the Bosumtwi crater, Ghana (Kontny et al., 2007). This observation gives rise to the question of whether iron-deficiency in pyrrhotite is a general feature associated with shock or a product of alteration. In any case the common feature must be a diffusion process producing an ordered structure responsible for the ferrimagnetic behavior. However, the abundant stacking faults distort such an ordering. This explains why solely the domains in between these defects are able to be structurally ordered. From this conclusion it can be furthermore derived, that the domains with an average diameter of 10 nm are large enough to exceed the superparamagnetic (SP) size range. Estimated pressures for the suevites in the CBIS in the range 10 - 35 GPa (Wittmann et al., 2009a) indicate that pyrrhotite has experienced some kind of shock deformation. Usually, pressure release after shock occurs in extreme short time intervals (e.g. Melosh and Ivanov, 1999; Stöffler et al., 2006), indicating that high energies are involved in these processes. Indeed, decompression creates the main deformation features in shocked minerals (Langenhorst, 1994; Stöffler et al., 1991) and can lead to a lower density of the post-shock phase compared to the pre-shock phase (Langenhorst, 1994) due to a rapid volume increase. The abundant lattice defects in iron-deficient pyrrhotite distort the close-packed ordering and thus evidence that the density of this phase is reduced. This is also confirmed by the Fe-deficiency, which is in accordance with an increased frequency of vacancies. Therefore, if the composition of iron-deficient pyrrhotite is related to shock, the iron loss must have occurred mainly during decompression. The large decompression energies then have triggered the motion of Fe ions out of the crystal. This process may have been favored by initial vacancies which are present in regular 4C structure. Partial vacancy reordering occurring subsequent to this process has then produced an ordered structure in small domains. If the iron-loss occurred during alteration rather than during decompression, the basic question is whether the iron-deficiency was already present prior to shock, or formed after shock. As discussed above, the transformation to the smythite-like structure is a result of LT oxidation in both cases. LT oxidation after shock is in accordance with the formation of secondary marcasite/pyrite. The latter are free of defects, evidencing a post-impact origin. Diffusion of iron and the establishment of an ordered structure are then a result of diffusion processes. However, the homogeneous chemical composition of the particular grains and the abundant defects question this interpretation. If diffusion has pervasively affected the pyrrhotite grains and produced a homogeneous chemistry, why are the abundant defects still present? In contrast, the interpretation stating that iron-deficient pyrrhotite was already present in the pre-shock rocks is in good accordance with the recent results. In this case, the defects have been introduced by shock and remained in the crystal. Post-impact alteration affected only parts of the grains and produced pyrite and marcasite, whereas other grain parts, which are now observed as iron-deficient pyrrhotite, remain rather unaffected by this process.

2.3.4 Conclusions

It has been shown that two types of pyrrhotite occur in the CBIS. The first is moderately altered, ductile and brittle deformed and shows no evidence of shock treatment higher than brittle deformation. The second type is shocked and strongly altered. Whereas unshocked pyrrhotite occurs exclusively in the schist block, both types are present in the suevite. Unshocked pyrrhotite is slightly depleted in iron and its chemistry is in agreement with "anomalous pyrrhotite" defined by Clark (1966). This pyrrhotite type originates from LT oxidation. The pervasive fractures of the grains in the sample from the CBIS make it very likely that anomalous pyrrhotite formed as a consequence of post-impact alteration. On the other hand, the formation of iron-deficient pyrrhotite is harder to explain. This pyrrhotite type is altered more strongly, but large amounts of internal defects indicate shock deformation. T_C is increased by about 25 - 40°C and the metal/sulfur ratio of 0.81 is in agreement with the composition of smythite. Collapse of the iron-deficient ferrimagnetic structure above 300°C not only shows that post-impact heating of the suevite must have stayed below that temperature, but equally indicates that this structure is favorably destroyed rather than back-transformed to the 4C structure. In the more general sense, iron-deficient pyrrhotite is the only stable magnetic sulfide phase in the suevite. This can be deduced from the fact that transitions indicating 4C and NC pyrrhotite are missing in the χ-T curves. It is not clear which mechanism induced the transformation from pyrrotite to iron-defcient pyrrhotite, since the homogeneous chemical composition indicating a pervasive transformation from pyrrhotite into a smythite-like phase conflict with the abundant lattice defects. If the transformation happened after the impact, the latter should have been annealed during reordering of the atomic structure. Therefore different scenarios can be proposed. The main question hereby is whether the transformation was mainly triggered by effects resulting from decompression after shock or by alteration. If decompression was the crucial driving mechanism, the structural vacancies and the structural anisotropy may have been important factors. It is well-established that pyrrhotite is distinctly less compressible parallel to the c-axis then along other dimensions (Kübler, 1985). Decompression should be therefore stronger perpendicular to c and generate defects favorably along the basal planes, which are partly or fully occupied with Fe-ions. The anisotropic crystal structure of pyrrhotite, consisting of a stacked sequences of basal planes along c, also favors vacancy movements parallel to the basis planes, indicating that vacancy reordering preferably occurred within Fe-layers parallel to c. This is in agreement with the defect-free domains between the stacking faults aligned parallel to the hexagonal (001) direction, which is perpendicular to the basal plane. The reordering process could have been triggered by a) internal energies, which remained in the crystal after shock and mainly arise from an unstable ordering, producing electrostatical energies, and residual elevated temperatures or b) alteration, respectively oxidation and associated diffusion processes. On the basis of the results presented above, however, it seems more likely that iron-deficient pyrrhotite was already formed before the impact event, even though no units containing such phases are known. This scenario is in agreement with both the homogenous chemical composition that was formed before the impact and the abundant defects, which have been introduced after the smythite-like structure was established. In this case the iron-deficiency is entirely independent of shock because pre-impact alteration has produced smythite. The fact that the measured c-lattice constant is not entirely equivalent to the one of smythite can then be explained by shock deformation, which slightly distorted the unit cell.

It seems clear that both shock- and alteration-induced processes modified the pyrrhotite structure. The present properties of iron-deficient pyrrhotite do not allow to clearly reconstruct its formation history. To shed more light on this problem, the last section of this work will present the results of an experimental setup, within which pyrrhotite has been exposed to definite shock pressures. The main purpose of these experiments was to study unmodified shock-related microstructures and magnetic properties of pyrrhotite in order to separate shock-induced modifications from those associated with alteration. These experiments were expected to give new information regarding the interpretation of iron-deficient pyrrhotite described in this section.

2.4 Shock experiments with pressures between 3 and 30 GPa on hexagonal and monoclinic pyrrhotite

Abstract

This section describes the results of shock experiments that have been conducted on a natural ore containing monoclinic (4C) and hexagonal (NC) pyrrhotite. The experiments have been designed on the basic problem appearing from the discussion in chapter 2.3. In that section it appeared that shock-, and alteration-triggered modifications in naturally shocked pyrrhotite are not clearly separable. A reasonable approach in order to shed light on this problem, however, can be achieved by building an experimental setup which allows to exclusively study primary shock features of pyrrhotite. Shock pressures in these experiments ranged from 3 - 30 GPa and the post-shock samples were subjected to microstructural and rock magnetic inverstigations similar as it has been done for the iron-deficient pyrrhotite in chapter 2.3. The experiments revealed a row of unexpected features, which can be roughly assigned to two different pressure regimes: Up to 8 GPa, microstructures in shocked pyrrhotite are characterized by mechanical deformation producing a damage of the crystal structure, whereas from pressures of 20 GPa upward, amorphization is the dominant shock-induced feature. The lower pressure regime is mainly characterized by an increase in coercivity, saturation isothermal remanent magnetization and coercivity of remanence with increasing pressure. These observations are in agreement with more single-domain (SD)-like behavior. Simultaneously, the peak associated with the λ-transition of NC pyrrhotite decreases and the 34 K transition of 4C pyrrhotite is broadened and depressed. Besides grain size reduction, which is generally assumed as the main driving mechanism for magnetic hardening, formation of SD within discrete multi domain (MD) grains presumably play another important role. PDFs subdivide such grains into lath-shaped domains with average sizes lying in the SD range. These PDFs disappear at 20 GPa and irregular, nm-sized amorphous domains occur instead. Pressure release from 20 GPa triggers segregation of native iron and at 30 GPa shock melting occurs. The latter is associated with crystallization of native iron and displays sharp interfaces to crystalline grain parts. The presence of native iron strongly influences the magnetic properties, depending on the particular amount in the studied sample and likely affects the magnetic properties of impact lithologies on Earth and extraterrestrial material.

2.4.1 Introduction

Monoclinic pyrrhotite, Fe_7S_8, is assumed to play an important role for the magnetization of meteorites (Rochette et al., 2001; Weiss et al., 2004), impact craters on Earth (Henkel., 1992; Kontny et al., 2007; Osinski et al., 2004) and Mars (Kletetschka et al., 2000; Louzada et al., 2007; Rochette et al., 2001).

Numerous experimental studies have been carried out in order to characterize the behavior of pyrrhotite under pressure (Ahrens, 1979; Gattacecca et al., 2007; Gilder et al., 2011; Kamimura, 1992; Louzada et al., 2010; Rochette et al., 2003), but the combination of magnetic and microstructural measurements remain rather untouched (Kontny et al., 2007; Mang et al., 2011). Especially during the last years the research on shock features in pyrrhotite has experienced an upturn and contributed fundamental new insights into the modifications of magnetic properties during and after shock. Basic features of the post-shock samples are an increase of the low temperature memory, bulk coercivity and saturation isothermal remanent magnetization (SIRM) (Louzada et al., 2007). These changes are mainly attributed to the fragmentation of large multidomain (MD) grains into smaller single domain (SD) grains (Gilder et al., 2004; Louzada et al., 2005) due to mechanical brecciation (Williamson et al., 1986). Under hydrostatic pressure, a ferrimagnetic to paramagnetic transition of Fe_7S_8 is reported that takes place at about 2.8 GPa (Rochette et al., 2003) or at ~6.8 GPa (Kamimura et al., 1992; Kobayashi et al., 1997). A high pressure phase (hpp) of pyrrhotite with a higher density (5.54 g/cm^3) compared to the low pressure phase (lpp: 4.93 g/cm^3) has been described by Ahrens (1979). Its formation is supposed to start between 2.7 and 3.8 GPa in small fractions, which then successively increase until the transformation is completed at ~25 GPa. The hpp is described to remain stable after shock (Ahrens, 1979). Kamimura (1992) found a four times lower compressibility along the c-axis around 6.8 GPa and suggested a collapse of the magnetic moments under pressure. Gilder et al. (2011) confirmatively calculated from direct measurements on compressed pyrrhotite that hysteresis loops become progressively broader until they collapse close to this transition. Whereas static compression leads to total demagnetization of the initial remanence above 2.8 GPa (Bezaeva et al., 2010; Rochette et al., 2003), dynamic pressure experiments show that pyrrhotite retains a minimum of 10% of the initial remanence after suffering shock pressures up to 12 GPa (Louzada et al., 2010). From his density determinations on shocked pyrrhotite, Ahrens (1979) already suggested that upon pressure release from above 25 GPa incongruent melting occurs. Shock-recovery experiments up to 60 GPa on the H6 chondrite Kernouvé, which contains troilite (FeS), indeed revealed localized shock-induced melting >30 GPa with FeNi metal and troilite (Schmitt, 2000).

In contrast to the magnetic behavior, shock-induced microstructural features of pyrrhotite are poorly documented (Ahrens, 1979; Kontny et al., 2007; Louzada et al., 2010). Microstructural shock features from a range of silicates allow estimating shock pressures for natural impact craters (French, 1998; Reimold and Koeberl, 2008; Stöffler et al., 1991). Many of such features have first been discovered in shocked quartz grains and are therefore best described for this mineral. With increasing pressure mosaicism fractures the initial grains and progressively planar fractures (PF), planar deformation features (PDF), diaplectic glass and high pressure polymorphs form (e.g. Fritz et al., 2011; Grieve et al., 1996; Stöffler and Langehorst, 1994; Stöffler et al., 1991). Diaplectic glass is mostly observed within PDFs (Ashworth and Schneider, 1985; Kieffer et al., 1976) and has a short range order (SRO). SRO phases do not show a crystal structure in the transmission electron microscope (TEM). Shock effects in troilite were investigated by Schmitt (2000) using reflected light microscope. The following microstructures are documented: undulatory extinction up to 25 GPa, twinning up to 45 GPa, partial recrystallization from 30 to 60 GPa, and complete recrystallization >35 GPa.

2.4.2 Material and experimental setup

The pyrrhotite ore

The starting material was a naturally deformed pyrrhotite ore from the Cerro de Pasco Mine, Peru. Its grain size varies between 0.01 and 1.5 mm and deformation is noticeable by the presence of subgrain structures and kinks, but no twins occur (Niederschlag and Siemes, 1996 and own studies, Fig. 2.4.1a). Randomly oriented cracks pervade the polycrystalline texture, which has been affected by ductile defor-

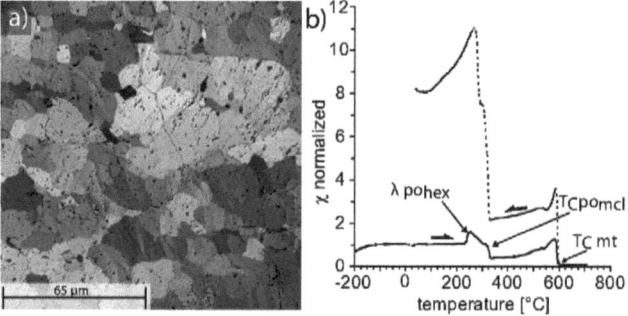

Figure 2.4.1: Characteristic features of the unshocked pyrrhotite ore from Cerro de Pasco, Peru. a: Reflected light with crossed polars. Undulous extinction (upper right corner), subgrains (lower central part), recrystallized grains (lower left) and cracks (central part) are present. b: Susceptibility (χ) as a function of temperature up to 700°C. Monoclinic (Curie temperature, T_{Cpomcl}: 325°C) and hexagonal (λ-transition, λ_{pohex}: 225°C) pyrrhotite, as well as magnetite (T_{Cmt}: 580°C) are identified. Pyrrhotite transforms partially into magnetite during heating, and hexagonal pyrrhotite is no longer stable in the cooling run.

mation. Magneto-optical imaging revealed that this ore mainly consists of hexagonal pyrrhotite. The monoclinic modification is restricted to cracks and grain boundaries. Such weakness zones break up first when stress is induced. Therefore, monoclinic pyrrhotite occurs mainly at grain rims of the broken fragments when shock has been applied. Magnetic lamellae, which reflect a typical twinning structure of monoclinic pyrrhotite (e.g. Kontny et al., 2000; Pósfai et al., 2000), become visible by coating the thin section with a magnetic ferrofluid. The hexagonal modification yields an average composition of 47.3 +/- 0.23 at.% Fe and corresponds to the NC modification (e.g. Desborough and Carpenter, 1965; Nakazawa and Morimoto, 1971). According to Niederschlag and Siemes (1996), the monoclinic phase contains 46.75 +/- 0.29 at.% Fe, which is consistent with the composition of monoclinic pyrrhotite described in literature (e.g. Arnold, 1962). Minor phases are galena, sphalerite, chalcopyrite, pyrite, iron oxides and gangue minerals. Both hexagonal and monoclinic pyrrhotite were identified by high-temperature magnetic susceptibility measurements (Fig. 2.4.1b). Upon heating, a peak (λ-transition) occurs at~240°C and displays the transition from antiferromagnetic NC to ferrimagnetic NA pyrrhotite (A and C are the axial length of the NiAs unit cell, N is an integer) (e.g. Haraldsen, 1937; Li and Franzen, 1996; Morimoto et al., 1975). The sharp drop at 325°C is related to T_C of ferrimagnetic 4C pyrrhotite above which the magnetic behavior is paramagnetic.

Experimental procedures

Randomly oriented pyrrhotite samples were cut into 15 mm diameter disks of 0.5 mm thickness and then subjected to defined shock pressures of 3, 5, 8, 20 and 30 GPa. The shock reverberation experiments with samples enclosed in a high impedance container (ARMCO iron) were conducted at the Ernst Mach institute in Freiburg using an experimental set up described in detail by Fritz et al. (2011), Meyer et al. (2011), Müller and Hornemann (1969) and Stöffler and Langenhorst (1994). An air gun and high explosives were used to accelerate a flyer plate with a given velocity onto the iron container in order to drive a well-defined shock wave into the ARMCO iron container. The shock pressure generated in the ARMCO iron is then enforced onto the sample by a series of reflection at the interfaces between the iron container and the pyrrhotite disk. The reverberation technique allows to subject well-defined shock pressures to geomaterial with less well known physical properties. Details for the different experiments are given in Table 2.4.1.

Table 2.4.1: Details of the shock experiments

sample	pressure p [GPa]	cover plate Ø [mm]	flyer plate Ø [mm]	acceleration tool	initial pressure iron [GPa]	velocity of flyer plate v [km/s]	thickness of sample [mm]
pyrrhotite	3	3	3	air gun	6	0.5	0.5
pyrrhotite	5	3	3	air gun	9	0.7	0.5
pyrrhotite	8	3	3	air gun	11	0.76	0.5
pyrrhotite	20	14.5	4	C4 (64) explosives	33	1.62	0.5
pyrrhotite	30	4	4	C4 (64) explosives	33	1.62	0.5

2.4.3 Results

After exposing five samples of the starting material to shock pressures of 3, 5, 8, 20 and 30 GPa, the samples were studied with respect to their magnetic behavior. Additionally, all samples were analyzed with optical and scanning electron microscopy; TEM was applied to selected samples.

Magnetic measurements

Besides pyrrhotite, accessory magnetite is present in some of the samples, but also forms during heating, which is especially true in some samples that suffered higher shock pressures. This is visible in the temperature-dependent susceptibility measurements (χ-T), which have been normalized to the initial value at room temperature for better comparison (Fig. 2.4.2). During heating, the magnetic susceptibility

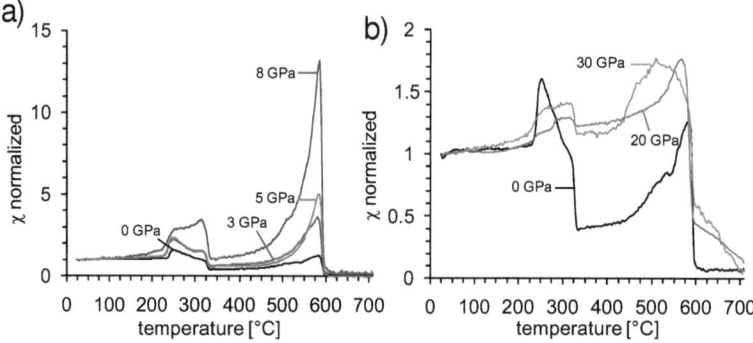

Figure 2.4.2: Susceptibility as a function of temperature (χ-T curve) for all samples. Only heating runs are shown. All curves are normalized to the initial value at room temperature. The pre-shock curve (0 GPa) is added as reference. a: 3 - 8 GPa. Between 0 and 5 GPa, the λ-transition appears as a peak around 245°C, which is constantly decreased upon warming down to~300°C and drops suddenly at 325°C (T_{Cpomcl}). In contrast, the susceptibility of the 8 sample remains rather constant above 245°C and slightly increases above 280°C, indicating the formation of monoclinic pyrrhotite during heating. b: 20 - 30 GPa. The relative contribution of monoclinic pyrrhotite to the susceptibility is significantly decreased, but the bulk susceptibility is significantly increased (see also Table 2.4.2). Above T_{Cmt} (580°C) the susceptibility is still enhanced, which is indicative for an iron phase.

rise at the λ-transition becomes larger with pressures up to 8 GPa (Fig. 2.4.2a), but is smaller for the 20 and 30 GPa samples (Fig. 2.4.2b). Generally, the susceptibility is enhanced between 220°C and 320°C and reveals that NC pyrrhotite is present in all samples and transformed to the NA type, even if a distinct peak is missing. The susceptibility of the samples shocked at 8 GPa and above is slightly temperature-dependent between 120 and 245°C, indicating that these phases lose stability during heating below the λ-transition. The latter appears as a sharp peak around 245°C followed by a subsequent constant curve decline up to~300°C at lower pressures, whereas the curves for 8 and 30 GPa in this temperature interval form a more plateau-like shape. Susceptibility tends to increases above 275°C and indicates that 4C pyrrhotite is formed during heating.

The χ-T curves of 8 - 30 GPa reveal that the initial fraction of monoclinic pyrrhotite is lower in this pressure range compared to the pre-shock sample, as the susceptibility at 325°C does not notably fall below the initial susceptibility level at room temperature. The susceptibility in the χ-T curves of 20 and 30 GPa has still not reached paramagnetic values beyond T_C of magnetite, but continuously decreases towards 700°C or and even beyond. This behavior suggests the occurrence of an iron phase, which has a Curie temperature of about 770°C (e.g. Garrick-Bethell and Weiss, 2010). Unfortunately, a second measurement up to 800°C could not re-prove this behavior because oxidation in air had already destroyed the iron phase. The NC structure is completely transformed to 4C pyrrhotite after heating up to 700°C in the post-shock samples (not shown in Fig. 2.4.2). The same behavior is already visible in the pre-shock samples (Fig. 2.4.1b).

In order to compare the stability of the NC phase from the pre- and post-shock samples, we performed a stepwise heating experiment (Fig. 2.4.3). The 0, 5, 8 and 30 GPa samples were heated to 200°C and cooled back while measuring the susceptibility. This experiment has then been repeated and the temperature was increased in steps of 20°C up to a final temperature of 340°C. All curves are reversible as long as the sample is heated below the onset of the λ-transition. Once this transition has been passed, the curve becomes irreversible. It is thus clear that the back transformation from NA to NC type on the cooling run is not completely reversible, which means that a fraction of NA pyrrhotite is still present as metastable phase due to an incomplete equilibrium reaction during cooling. During the repeated heating steps, the λ-transition starts at a slightly lower temperature and the slope of the susceptibility increase is less steep. If a new sample is used for each heating step, the λ-transition appears sharp and its slope remains steep in all runs. Pre- and post-shock samples behave similar (Fig. 2.4.3), but at higher shock pressures the raise in susceptibility is decreased. These experiments reveal that the NC-NA transition is still present but weakly hampered in all pre- and post-shock samples up to 30 GPa. Furthermore, T_C of the 4C phase at 325 °C remains reversible in all runs. However, compared to the pre-shock sample, the post-shock samples differ in their χ-T curves by less steep susceptibility slopes at the transition temperatures. This indicates a change in domain size from MD to SD grain size with increasing shock pressure.

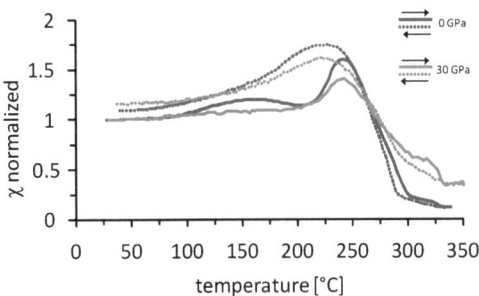

Figure 2.4.3: Susceptibility as a function of temperature (χ-T curve). The final runs of successive heating curves (up to 340°C) are displayed for 0 and 30 GPa, respectively. Both hexagonal and monoclinic pyrrhotite are stable during the heating experiments, but phase transitions are hampered in the shocked sample. For further explanations, see text.

Monoclinic 4C pyrrhotite undergoes a low temperature magnetic transition at 34 K (Rochette et al., 1990), corresponding to the transformation of monoclinic to triclinic structure (Wolfers et al., 2011). Therefore, pre- and post-shock samples were studied by low-temperature magnetic susceptibility and saturation isothermal remanence magnetization (SIRM) measurements. Low-temperature (LT) in-phase (χ') and out-of-phase (χ") susceptibility measurements are shown in Fig. 2.4.4. During warming, all samples show an increase in their in-phase susceptibility between 20 and 40 K, which is typical for

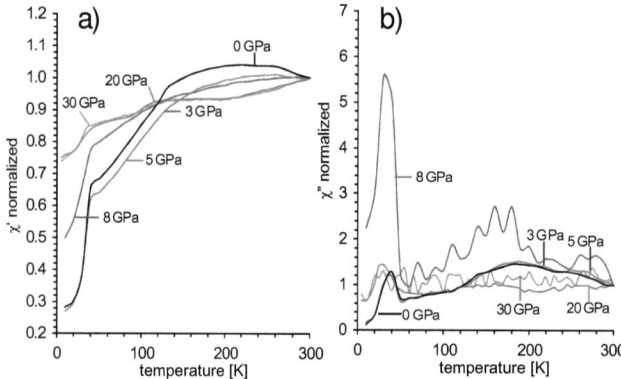

Figure 2.4.4: Susceptibility as a function of temperature below 300 K. a: in-phase susceptibility (χ'). The susceptibility generally becomes less temperature-dependent with increasing pressure. Equally, the 34 K transition decreases above 5 GPa and the curve is flattened below this temperature, especially between 20 and 30 GPa. b: Out-of-phase susceptibility (χ''). The strong peak at 34 K in the pre-shock sample has disappeared at 30 GPa.

monoclinic pyrrhotite. A less steep increase follows until ~120 K (the Verwey transition (T_V) of magnetite), and a further very slight increase until 300 K. The jump in susceptibility at the 34 K transition becomes less pronounced and smears out for the 20 and 30 GPa samples. The kink in the curve is shifted by about 8 K to lower temperatures in the 30 GPa sample. In the unshocked sample, the kink in the in-phase susceptibility coincides with a frequency-dependent peak in the out-of phase susceptibility (Fig. 2.4.4b) indicating magnetic relaxation. This peak increases in the 8 GPa sample, is shifted by about 10 K to lower temperatures in the 20 GPa sample and is absent in the 30 GPa sample.

Field cooled (FC) and zero field cooled (ZFC) LT SIRM show a significant drop at the LT-transition upon warming, which becomes less pronounced with increasing pressure (Fig. 2.4.5a) and is less steep above 8 GPa (Fig. 2.4.5b). The 30 GPa sample shows a second kink at about 15 K in addition to the 34 K transition. In the 20 GPa sample, a second drop of SIRM occurs at 130 K in the ZFC and at ~150 K in the FC curve. Approaching the LT-transition from above produces a different SIRM behavior in unshocked 4C pyrrhotite, similar to what is observed for MD magnetite at T_V (e.g. Jackson et al., 2011). During cooling from room temperature (RT), SIRM first gradually decreases until a first kink appears at 120 K, which is the T_V of magnetite. T_V is also visible in the 8 GPa sample and indicates minor amounts of magnetite. Upon further cooling RTSIRM slowly decreases and shows a significant loss of remanence at the LT-transition of monoclinic pyrrhotite. During warming, a major remanence recovery occurs at the transition (Table 2.4.2) and the remanence remains roughly temperature-independent until 300 K. However, 18% of the remanence are lost after cycling through the LT transition, which is in agreement with a MD-PSD character of unshocked pyrrhotite at room temperature (see Dekkers et al., 1989 for calculations). The LT transition reversibility (LT_{rev}) increases significantly from 0 to 3 GPa and remains approximately equal for higher pressures (Table 2.4.2). An increase in LT_{rev} is in line with grain size reduction (Dekkers et al., 1989). Intensive brecciation occurs already at 3 GPa and is reflected by an increase of the LT_{rev}. The value of 0.88 gives average grain sizes of ~10 μm (Dekkers et al., 1989). Above

Table 2.4.2: Magnetic properties of pre- and post-shock pyrrhotite samples

p [GPa]	RT χ' [m^3/kg] $\cdot 10^{-5}$	RT χ'' [m^3/kg] $\cdot 10^{-7}$	M_s [Am^2/kg]	M_{rs} [Am^2/kg]	H_c [mT]	H_{cr} [mT]	M_{rs}/M_s	H_{cr}/H_c	LT$_{rev}$
0	0.55	0.47	2.03	0.42	22.71	35.45	0.21	1.56	0.82
3	0.33	0.04	1.74	0.64	41.09	62.46	0.37	1.52	0.88
5	0.46	0.25	1.93	0.55	32.44	48.47	0.29	1.49	0.86
8	0.27	-0.06	1.63	0.69	57.95	77.96	0.42	1.35	0.86
20	2301.58	1583.18	13.85	1.34	9.96	27.65	0.10	2.78	0.89
30	296.47	142.29	1.35	0.54	51.28	62.29	0.40	1.21	0.86

p: pressure; RT: room temperature; H$_{cr}$: coercivity of remanence; H$_c$: coercivity; M$_{rs}$: saturation remanent magnetization; M$_s$: saturation magnetization; LT$_{rev}$: degree of reversibility of the 34 K transition during a cooling-warming cycle (see Dekkers et al., 1989).

that pressure, the LT$_{rev}$ remains rather constant since grains with average diameters below 10 µm are accompanied by minor changes in the LT$_{rev}$, which lie within the analytical errors.

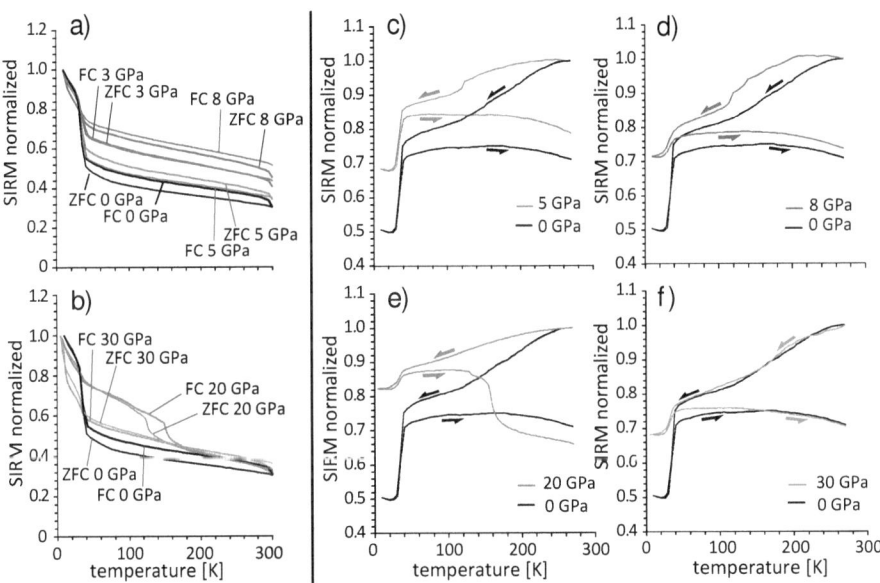

Figure 2.4.5: Saturation isothermal remanent magnetization (SIRM) as a function of temperature. a + b: Field cooled (FC) and zero field cooled (ZFC) measurements for all samples. The drop in remanence decreases with increasing pressure, except in the 30 GPa sample. c - f: Room temperature (RT) SIRM. Reversibility increases above the 34 K transition but decreases below that point from 0 to 20 GPa. The 30 GPa curve, in contrast, re-approaches to the one of the pre-shock sample. An additional drop in remanence occurs at 20 GPa between 130 and 150 K and is presumably caused by alabandite (Kohout et al., 2010).

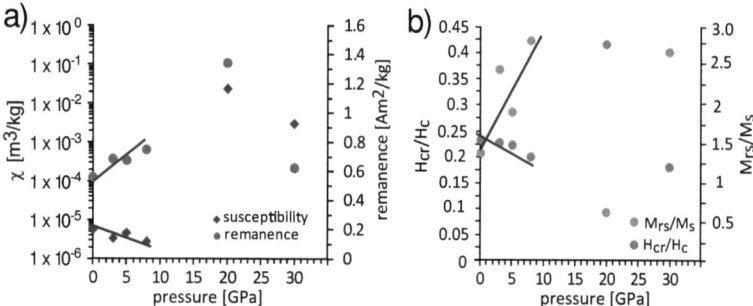

Figure 2.4.6: Selected magnetic parameters vs. pressure. a: Absolute susceptibility and remanence values. Trends (marked by trend lines) appear between 0 and 8 GPa but are not traceable above. The values of 20 and 30 GPa are additionally influenced by fractions of native iron. b: H_{cr}/H_c respectively M_{rs}/M_s vs. pressure. Trend lines can again be interpolated between 0 and 8 GPa.

The SIRM drop at the LT transition is strongly decreased above 5 GPa and so is the reversibility below this transition, especially at 8 GPa (Fig. 2.4.5d). The 20 GPa sample shows an additional drop in SIRM during warming (Fig. 2.4.5b + e) at~130 K. This feature is observable only after cycling through the LT-transition of monoclinic pyrrhotite and is most likely related to the presence of an additional magnetic phase. The presence of such a phase would inhibit reasonable determinations of the LT memory, as calculated from the initial and final SIRM value in the RTSIRM curves. The same holds for magnetite, which is also present in some of the samples. Therefore this parameter is not considered here. Although the LT-transition is still small in the 30 GPa sample, the SIRM drop and the reversibility below 34 K is again increasing, indicating larger magnetic domain sizes. Fig. 2.4.6a shows RT magnetic susceptibility and SIRM versus shock pressure. Up to 8 GPa, a clear trend is visible following the decrease of magnetic susceptibility and the increase of SIRM. The 20 and 30 GPa samples are distinctly different. While both samples show a significant increase in magnetic susceptibility, the SIRM in the 20 GPa sample is about twice as high as in the 30 GPa sample. The reason for this behavior is the formation of an iron phase in the 20 and 30 GPa samples, which is in agreement with the HT χ-T curves. Trends between pre-shock and the 8 GPa sample equally appear in the hysteresis parameters (Fig. 2.4.6b, Table 2.4.2). These trends basically reveal an increase in coercivity (H_c) with increasing pressure, except the sample at 3 GPa which has a higher H_c than the 5 GPa sample. The same behavior is observed for the saturation remanence. The H_{cr}/H_c (H_{cr}: remanence of coercivity) ratio decreases from 1.56 (0 GPa) to 1.35 (8 GPa) and the ratio between remanent saturation magnetization to saturation magnetization (M_{rs}/M_s) increases from 0.21 to 0.43 (Fig. 2.4.6b) in agreement with decreasing grain sizes. The 20 GPa sample has an extremely low H_c (~10 mT) and M_{rs}/M_s ratio but a conspicuously high H_{cr}/H_c ratio. This sample therefore stets itself clearly apart from all other samples. In deed, hysteresis parameters of the 30 GPa sample are again close to those of the 8 GPa sample, even if all except M_s are comparatively decreased. Leaving apart the 20 GPa sample, a downward trend is visible between 0 and 30 GPa in the H_{rc}/H_c ratio with a flattened slope above 8 GPa (Fig. 2.4.6b).

Microstructures

The most obvious microscopic feature appearing with increasing shock pressure is a non-linear increase in brecciation and the formation of micro-cracks within the remaining grains. A higher degree of brecciation is accompanied by a larger range of observed grain sizes and a higher fraction of ultrafine grains. Brittle deformation affects all samples up to 30 GPa. However, the percentage of angular fragments and internal microcracks is at maximum in the 20 GPa sample (Fig. 2.4.7b). The magnetic domains in the fer-

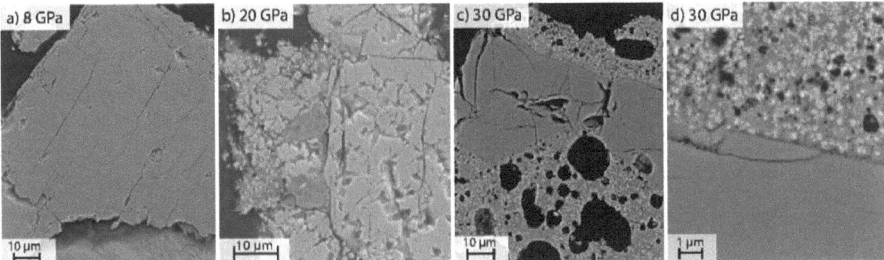

Figure 2.4.7: SEM photographs of shocked pyrrhotite. a: One set of parallel fractures is present in the 8 GPa sample, SE image. b: Brittle deformation strongly pervaded the sample at 20 GPa and produced abundant angular grains, BSE image. c + d: Melt is visible in the 30 GPa sample and forms a sharp interface to the crystalline material, BSE image. SEM: scanning electron microscope; BSE: backscattered electron; SE: secondary electron.

rimagnetic 4C pyrrhotite, which are visualized by a magnetic fluid (Fig. 2.4.8a, b), progressively vanish with increasing pressure. Instead of linear domain structures, the fluid accumulates at tiny spots, suggesting a destruction of the original magnetic lamellae structure (Fig. 2.4.8). The magnetic lamellae are suggested to mirror twin domains in the monoclinic pyrrhotite (e.g. Kontny et al., 2000; Zapletal, 1993). In the post-shock samples they first become repeatedly interrupted (Fig. 2.4.8d - e) until they disappear and the ferrofluid concentrates in clusters at the grain rims instead. The clusters mainly lack an internal structure and sometimes show "worm-like" shapes (Fig. 2.4.8f - g).

Microscopically, pyrrhotite shocked up to 20 GPa is affected by intense brittle deformation. Above 20 GPa, shock melting begins (Fig. 2.4.7c). Besides completely molten fragments, partially molten grains are present (Fig. 2.4.7c), wherein the interface between molten and unmolten areas in these grains is sharp and well-defined (Fig. 2.4.7c, d). Melt has evidently been rapidly quenched since the adjacent grain part is generally free of visible heat overprint. The vesicular melt contains abundant skeletal iron crystals of less than 0.5 µm in size (Fig. 2.4.7d) suggesting an incongruent melting of pyrrhotite. The formation of iron explains the high magnetic susceptibility and the thermomagnetic behavior in the 20 and 30 GPa samples.

TEM investigations allow a more detailed study of the deformation features that occur within the crystal lattice of shocked pyrrhotite. The frequency of small-scaled lattice defects such as dislocations and irregular bending of lattice planes is at maximum in the 8 GPa sample. This sample additionally contains a set of amorphous lamellae with a width of 5 - 60 nm (Fig. 2.4.9a, b). These lamellae run parallel to each other, are slightly bended and are interpreted as PDFs. On both sides of the observed lamellae, crystalline blocks have been broken off and rotated in one direction. This observation suggests that the formation of the lamellae is linked with a shear component. Such an interpretation is consistent with observations from Langenhorst and Greshake (1999) in PFs. Leroux et al. (1994) showed that shock-induced amorphous bands behave as ductile layers and thus act as small-scaled shear zones. Parallel to the PDFs, small-scaled bands, which indicate shear movements are indeed visible in high-resolution (HR) TEM (Fig. 2.4.9b). These bands are oriented parallel to each other and have an average distance of ~7 nm. Within such a band, the lattice planes seem to be bended into a sigmoidal structure, similar to ductilely deformed minerals in shear zones (e.g. Passchier and Trouw, 2005) and likely display small-scaled shear bands. The shear band fabric in the crystalline blocks at the rims of the PDFs rotated with the blocks, thus proving that the lamellae formed after the shear bands.

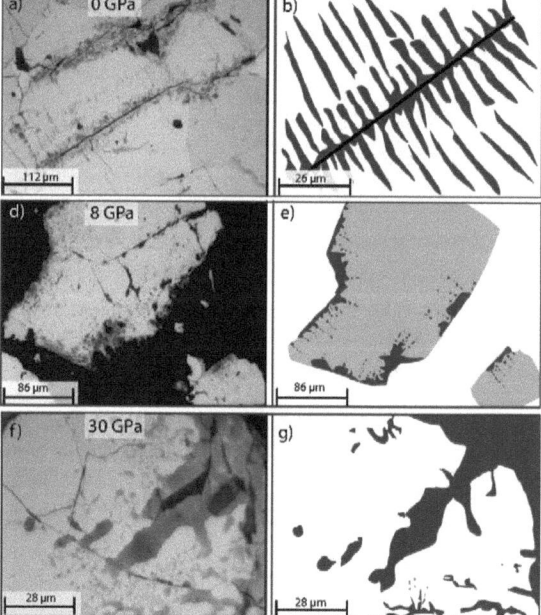

Figure 2.4.8: Microscopic photographs (left hand side) and corresponding schematized sketches (right hand side). The samples have been coated with a magnetic fluid, which sticks at the magnetic domains. The typical twinning structure of monoclinic pyrrhotite, which is visible in the pre-shock sample, disappears with increasing pressure.

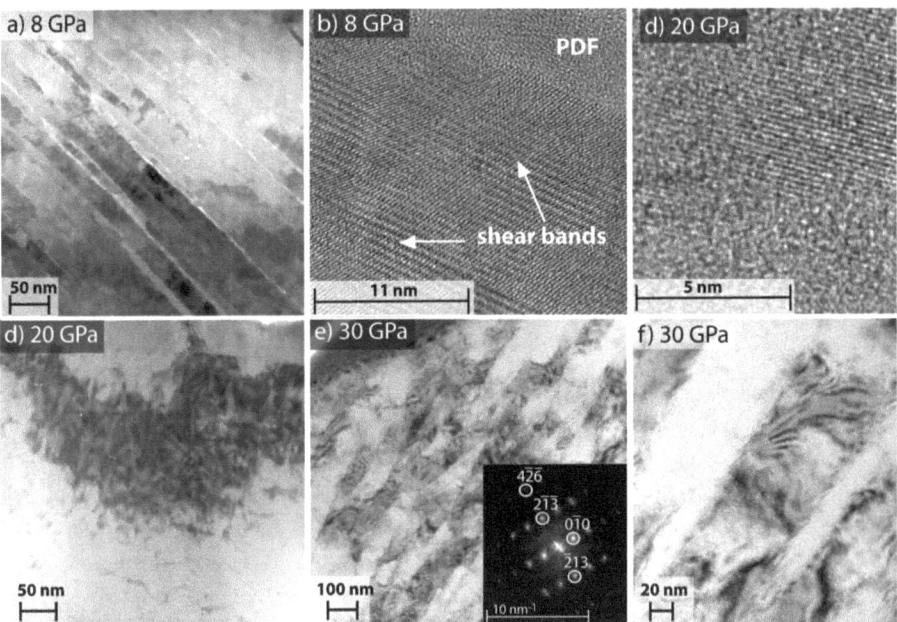

Figure 2.4.9: TEM photographs. a: Bright-field image illustrating planar deformation features (PDFs) (white parallel streaks). b: High resolution (HR) TEM image showing strong mechanical deformation of the crystal lattice. A detail of a PDF is visible to the upper right image part; these structures form amorphous lamellae. Small-scaled shear bands run parallel to the PDFs and indicate that a shear component was involved during lattice deformation. c: Bright-field (BF) image showing rippled Fe-accumulations that consist of Fe and S. The bulk composition distinctly varies within few nanometers and Fe has an unusual high percentage (between 52 and 58 at.%). d: HRTEM photograph of an interface between crystalline (image center) and amorphous (upper and lower image part) areas in the 20 GPa sample. The boundary between both is blurred and continuous. e + f: Bright field image showing discontinuous lath-shaped domains interpreted as deformation twins. The inset in e) shows a diffractogram of the 30 GPa sample. Reflections are characteristic for the monoclinic 4C structure. TEM: transmissions electron microscope.

The 20 and 30 GPa samples display a distinctly different internal microstructure. Shear bands are absent and the frequency of lattice defects is drastically decreased compared to the 8 GPa sample. The 20 and 30 GPa samples show a distinctly higher amount of defects compared to the crystal lattice in the pre-shock sample, but compared to the 8 GPa sample the crystal lattice is intact. However, this is only true for some regions of the crystal. HRTEM reveals that nm-sized lattice areas in these samples lack of any crystal structure and are amorphous (Fig. 2.4.9d). In such areas the typical contrast for lattice fringe imaging is missing. They occur randomly within the crystal lattice and increase in size and frequency from 20 to 30 GPa. The interface between amorphous and "crystalline" parts is not sharp but appears gradually (Fig. 2.4.9d).

A distinct feature in the 20 GPa sample is the occurrence of diffuse patches at few places in the sample (Fig. 2.4.9c). These structural details appear in TEM bright-field (BF) images as irregular, darker areas in which lighter and darker areas alternate. The internal structure of such areas often appears as rippled and diffuse stripes. In TEM the electrons are generally more strongly scattered by elements with a high atomic number, which is the reason why areas containing such elements appear darker in BF images (given that the sample has a constant thickness). Differences in brightness can thus be indicative for variations in the chemical composition which is indeed true for this sample. EDX analyses on several points within these "rippled" areas give values varying between 53 and 58 at.% for iron. This compositional range is above that of pyrrhotite or even troilite, and therefore suggests an intergrowth of pure iron with a Fe-S phase. To avoid confusion with the amorphous domains described above, these domains will be named "rippled Fe-accumulations" in the following.

Lath-shaped domains with an average spacing of 50 nm are a prominent and exclusive feature in the 30 GPa sample and represent deformation twins (Fig. 2.4.9e - f). The dark-bright pattern occuring within one set of domains in Fig. 2.4.9e and f either indicate a strong internal stress field or Moiré pattern with interfering tiny crystallites. The latter would be in agreement with recrystallization of the pyrrhotite. In any case these structures explain the behavior of the ferrofluid on the shocked samples (Fig. 2.4.8). The progressive distortion of the crystal lattice generates small magnetic domains which successively fall below the size that can be displayed by accumulation of the ferrofluid.

Table 2.4.3 resumes the crystallographic parameters of all studied TEM samples. The 20 GPa sample is excluded here, since it was not possible to fully observe the present zone axis and the results are

Table 2.4.3: Crystal parameters of pre- and post shock pyrrhotite samples studied with TEM

	0 GPa	8 GPa	30 GPa
a [Å]	5.744	6.867	6.867
b [Å]	3.446	6.867	6.867
c [Å]	5.976	17.062	17.062
α	90°	90°	90°
β	90°	90°	90°
γ	90°	120°	120°
crystal system	orthorhombic	trigonal	trigonal
modification	5C	4C	4C
chemical formula	$Fe_{0.91}S$	$Fe_{0.875}S$	$Fe_{0.875}S$

therefore not well defined. Regarding the parameters especially at 30 GPa, it becomes immediately clear that shock up to 30 GPa does not form a new hpp. The lattice parameters of the 8 and 30 GPa are consistent with the 4C modification of pyrrhotite and have not been modified. Microprobe analyses, which reflect average composition of larger grain areas (average Ø: 1 μm), are in agreement with these results, showing typical average compositions of NC pyrrhotite for all shock stages (Table 2.4.4).

Table 2.4.4: Average compositions of pre- and post-shock pyrrhotite samples

	0 GPa		5 Gpa		8 GPa		20 GPa		30 Gpa	
atomic ratio	S	**Fe**	S	**Fe**	S	**Fe**	S	**Fe**	S	**Fe**
mean value	52.44	47.54	52.82	47.1	52.48	47.45	52.33	47.60	52.39	47.58
st. dev	(010)	(010)	(042)	(043)	(024)	(024)	(021)	(021)	(014)	(012)
n	32		35		45		42		43	

st. dev.: standard deviation

2.4.4 Discussion

Permanent shock-induced microstructural peculiarities influence the intrinsic magnetic properties of pyrrhotite. It could be shown that trends revealing an increase of SIRM and a decrease of the susceptibility occur in the samples shocked up to 8 GPa. The microstructural studies showed that this pressure range is identical with the one within which mechanical deformation determines the formation of particular microstructural features. Despite the fact that the magnetic properties of the 20 GPa sample may be influenced by a second magnetic phase, our results clearly show the appearance of new deformation mechanisms at 20 GPa, which are dominated by diffusion. The latter produces amorphization and melting of the crystalline material. These two pressure regimes can thus be distinguished by the occurrence of the particular residual deformation structures in the post-shock samples (Table 2.4.5). These observations might be critical for the explanation of magnetic properties of impact-related lithologies on Earth and meteorites.

3 - 8 GPa regime: internal stress and fracturing in pyrrhotite

The increase of coercivity and saturation remanence, which goes along with a decreasing susceptibility upon increasing pressure (Fig. 2.4.6 and Table 2.4.2), is in agreement with observations of earlier studies (Louzada et al., 2007, 2010). This behavior is consistent with a progressive grain size reduction due to brittle fracturing and is the most obvious deformation mechanism in this pressure range. Although the main difference in grain size reduction occurs between the pre-shock sample and 3 GPa (Table 2.4.2), approximately linear trends appear for most parameters within this pressure regime. An increase of the M_{rs}/M_s ratio appears in the post-shock samples from 0 to 8 GPa. This feature is known to develop already while pressure is applied (Gilder et al., 2011). Magnetic parameters, however, are generally enhanced in the 3 GPa sample compared to 5 GPa. One has to remember that the former shock pressure lies near the Hugoniot-Elastic limit for pyrrhotite (3 - 3.5 GPa) (Louzada et al., 2010 supplemental). Brittle deformation in the 3 GPa sample clearly denotes permanent deformation, but in particular also elastic deformation, which should compensate the main stress at this pressure. This deformation may have left strong internal stresses after shock since the pressure was already very high. Internal stress can change the magnetostriction constants (Gilder et al., 2004) or inhibit domain wall movement. At 5 GPa, the Hugoniot-Elastic limit has been overpassed and stress could then be fully compensated by internal strain producing permanent deformation structures as seen in Fig. 2.4.8 and 2.4.9a, b. High internal stresses produced near the Hugoniot-Elastic limit likely affect the magnetic properties stronger than internal deformation produced in a somewhat higher pressure stage (5 GPa). Increasing brecciation results in a larger fraction of grains with SD-like behavior, which is also reflected in the LT reversibility (Table 2.4.2). The pre-shock samples show a LT reversibility of 0.80 suggesting average grain sizes of about 15 - 20 μm (Fig. 2 in Dekkers et al., 1989). This is in good agreement with the microscopic magnetic domain observations of this study depicted in Fig. 2.4.7a. The relatively small grain sizes of monoclinic pyrrhotite in the starting material explain the fact that differences in the magnetic parameters between pre- and post-shock samples are not as strong as observed in other studies (Louzada et al., 2007). Following Hodych (1990), the slope of LT hysteresis parameters in a plot where the M_s is plotted against

Table 2.4.5: Summary of principal magnetic and microstructural features in shocked pyrrhotite

pressure [GPa]	magnetic features	microstructures	connection between features
0-8	• decrease of λ-transition • broadening and depression of 34 K transition • increase of H_c, H_{cr}, M_s and LT_{rev}	• brecciation (grain size reduction) • nano-scaled shear bands • one set of PDFs	• amorphous lamellae divide PSD grains into elongated SD grains • defects enhance domain wall pinning
20-30	• remanence and susceptibility become less temperature-dependent • M_s, M_{rs} and SIRM decrease compared to 8 GPa (20 GPa sample excluded)	• no PDFs or nano-shear bands • rippled Fe-accumulations at 20 GPa • amorphous domains with increasing frequency from 20 to 30 GPa • melt with crystallized native iron • deformation twins at 30 GPa • crystalline pyrrhotite consists of 4C and NC pyrrhotite	• native iron additionally modifies magnetic parameters • amorphous domains reduce the fraction of domains which can potentially acquire a SIRM • deformation twins enhance SP fraction • hpp absent in the crystalline grain parts • amorphous domains are remnants of the hpp or of diaplectic glass

H_c clarifies the present magnetic domain state. Fig. 2.4.10 illustrates that a distinct fraction of PSD grains is already present in the pre-shock sample, as the trend line crosses the y-axis explicitly above zero due to enhanced coercivities. The 8 GPa trend line reveals even higher coercivities, which are in agreement with a single domain grain fraction with shape anisotropy (Hodych, 1990). A similar conclusion can be drawn from the relatively high M_{rs}/M_s ratio (0.42) at room temperature for this sample because a pure assemblage of single domain grains with shape anisotropy should produce M_{rs}/M_s ratios of ~0.5 (Dunlop, 2002). Microstructural images of this study (Fig. 2.4.8a, 2.4.9a) reveal that lattice preferred fractures and PDFs pervade large parts of the crystal lattice at 8 GPa, defining elongated magnetic domains. These domains interrupt the previous magnetic domain structures and most likely force the latter to re-arrange within the predetermined frame. Such frames have a high aspect ratio and are small enough to produce SD behavior (e.g. Dunlop and Özdemir, 1997). Furthermore, the magnetic moments should arrange parallel to the long axis (Jackson, 1991), which in return causes a uniform magnetization within a particular grain since the frames are oriented parallel to each other. Domain re-structuring as an explanation of the observed increase in SD behavior in shocked magnetic minerals has already been suggested in previous works (Gilder et al., 2004; Louzada, 2008).

The increase of total SIRM, H_c and M_{rs}/M_s ratio upon pressure (Fig. 2.4.9b, Table 2.4.2) is consistent with changes in the magnetostriction and magnetocrystalline constants (Gilder and LeGoff, 2002; Gilder et al., 2004). Defects and residual strain lead to stress hardening which stabilizes the present atomic ordering. This effect should also influence the 34 K transition, which is indeed hampered at 8 GPa. Because this transition is related to structural modifications (Wolfers et al., 2011), it is likely that defects depress and broaden the transition similarly to how it has been observed for non-stoichiometric magnetite at the Verwey transition (Aragón et al., 1985).

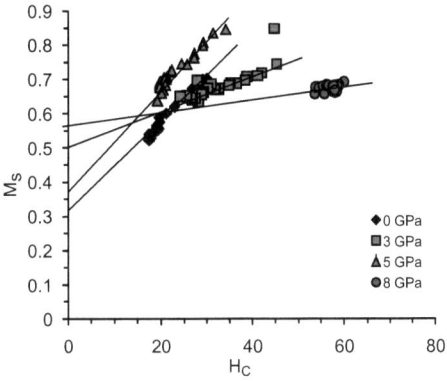

Figure 2.4.10: Saturation magnetization (M_s) vs. coercivity (H_c) measured during LT hysteresis measurements between 10 and 300 K. These measurements have been done for the samples shocked up to 8 GPa. Linear trends in this plot indicate that the pre-shock sample already contains large fractions of PSD grains. The flat slopes of these trend lines can be explained by a fraction of SD grains with shape anisotropy (Hodych, 1990).

20 - 30 GPa regime: partial amorphization, melting and crystallization of iron

The onset of amorphization in the 20 GPa sample is coupled to the disappearance of lattice defects and amorphous lamellae. Also, melting and crystallization of iron occur in this pressure range and significantly change the rock magnetic properties (Table 2.4.2). Especially the 20 GPa sample is affected by some peculiarities causing deviations from all trends in the magnetic measurements. As described below, the starting material of this sample was a pyrrhotite ore that contained minor phases and it cannot be excluded that some of these phases are present in the shocked samples. Most of these phases are paramagnetic at room temperature except of magnetite which is recognizable by a slight drop in RTSIRM in the 5 and 8 GPa samples (Fig. 2.4.5c, d). The remanence drop at~150 K in the 20 GPa sample (2.4.5 a, b, e), however, is distinctly different and additionally appears in the ZFC and FC curves. Therefore magnetite is excluded to be the source of that behavior. Alabandite, (Fe,Mn)S is a much more likely candidate since Mn is common within iron sulfide ores. The formation of alabandite under the hydrothermal conditions that governed the original ore formation would thus not be surprising. This mineral is paramagnetic at room temperature, but has two transitions at 130 and 148 K. The former is described as a phase transition (Kohout et al., 2010) and the latter is the Néel temperature below which alabandite becomes ferrimagnetic (Heikens et al., 1977; Kohout, 2010). This behavior is consistent with the remanence drop upon warming and suggests that the remanence behavior of this mineral is strongly dependent on the temperature at which the SIRM was given, similarly as it is the case for magnetite (e.g. Özdemir et al., 2002). This phase, however, does not explain the striking magnetic properties of the 20 GPa sample at room temperature. The hysteresis parameters (Table 2.4.2) suggest that a large amount of iron dominates this sample. On the one hand, larger melt droplets were not detected in the SEM images of this sample. On the other hand, however, TEM analyses and the χ-T curves (Fig. 2.4.2b) suggest the presence of native iron. The M_{rs}/M_s ratio of 0.096 (Table 2.4.2) approximately fits with the hysteresis parameters given for pure iron by Gose (1971). In addition, the relatively low H_c and high M_s value indicate the dominance of iron (see Table 3.1 in Dunlop and Özdemr, 1997). But why is the dominance of native iron so high in this sample athough no Fe-particles where detected on the µm-scale? A likely explanation is contamination of iron by the iron container, which was used for the experimental setup. Even though the sample was carefully removed from the container with a non-metallic subject, it can not be fully excluded that small flakes of iron have been peeled off the container rim. At 20 GPa, injection of molten iron from the container into the sample is equally a possible process, but in any case, the large dominance of iron make this sample unsuited for further discussion on shocked pyrrhotite. The discussion about microstructures and magnetic properties of pyrrhotite shocked within the higher pressure regime is therefore going to be focused on the 30 GPa sample. Although brecciation is even stronger at 30 GPa, M_{rs}/M_s and H_{cr}/H_c ratio are reduced compared to the 8 GPa sample and so are H_c and M_s. This observation argues for a decrease in SD-like behavior in the 30 GPa sample, which can be related to the missing PDFs and the associated subdivision of larger grains. However, deformation twins should produce a similar effect, but here the relatively small distances between boundary spacing is suggested to produce an additional fraction of domains with SP behavior. Since the latter contributes to M_s but not to M_{rs} (Özdemir et al., 2002), this fraction lowers the M_{rs}/M_s ratio. The total reduction of M_s for 30 GPa compared to 8 GPa indicates that the random occurrence of amorphous spots may hamper the acquisition of an imposed remanence.

Annealing and diaplectic glass

Regarding the distinct microstructral differences between 8 and 20 GPa, one must discuss the role of annealing within the upper pressure regime. In general, shock-induced heating is governed by the compressibility of the shocked material. Thus, overshooting the pressures required to physically crush the crystal lattice (i.e. transformation of crystalline quartz into diaplectic glass) produces a substantial increase of pressure-volume work and leads to deposition of energy by the passing shock wave (i.e. Fritz et al., 2001). Calculations of post-shock temperatures (Artemieva and Ivanov, 2004; Fritz et al., 2005) are limited due to the sparse existence of Hugoniot data for pyrrhotite (Louzada et al., 2011, supplemental). For comparison, shock loading of 30 GPa would result in a post-shock temperature increase of 80, 130 and 530°C for gabbro, iron and sandstone, respectively (Table 2 in Meyer et al., 2011). For several non-porous geo-materials, the estimated shock temperatures would be distinctly below the temperatures required to melt pyrrhotite (Jensen, 1942). Although shock-induced microstructures in metallic phases are particularly sensitive to temperature (e.g. Leroux, 2001), it is rather unlikely that annealing completely erased the complex deformation features observed at 8 GPa, but left unaffected other features as PDFs, rippled Fe-accumulations and deformation twins (Fig. 2.4.9c - f). Sharp boundaries between molten and crystalline pyrrhotite indicate quenching and suggest that peak temperatures persisted only over a limited time interval. However, the better pronounced transition at 34 K (Fig. 2.4.4a, 2.4.5f) indicates that temperatures were sufficient to roughly reorder smaller defects that strongly hamper the transition at 8 GPa. It is thus concluded that the samples shocked in the upper pressure range of the here presented experiments have been affected by minor annealing effects, mainly due to rapid cooling.

In shocked minerals two different processes are generally found to produce amorphous structures: melting, or at lower pressures the formation of diaplectic glass (e.g. Feldmann et al., 2006; Stöffler et al., 1991). The latter is seen as the direct transformation from crystalline to amorphous atomic ordering without melting. In most of the cases, when amorphous domains are found in shocked silicates (e.g. Ashworth and Schneider, 1985; Langenhorst 1994; Leroux et al., 1994), they are restricted to lamellar frames (Goltrant et al., 1991, 1992; Kitamura et al., 1977) similar to the PDFs in the 8 GPa sample of this study. Such lamellae structures are supposed to produce diaplectic glass (Langenhorst, 1994) by increasing width and frequency. Stöffler (1966) found sets of amorphous twin lamellae along with crystalline twin lamellae in shocked plagioclase crystals of the Nördlinger Ries crater. Amorphous structures in samples from the higher pressure regime of this study, however, occur in form of shapeless domains without sharp contact to the crystalline material (Fig. 2.4.9d) and are not restricted to any crystallographic feature. Fritz et al. (2005) found spots of maskelynite in shocked plagioclase under optical microscope and could show that different shock stages were present in one single grain. Occurrence of diaplectic glass (maskelynite) could not be restricted to specific crystallographic features in that study. Up to the knowledge of the author, however, a scattered occurrence of amorphous spots in the nm-scale has not yet been reported. If these amorphous spots represent diaplectic glass, such structures are formed in pyrrhotite by amorphization of random spots in the crystal lattice. The spots progressively increase in size and frequency with pressure and are then suggested to pervasively affect the crystal structure above 30 GPa.

Shock melting

The microstructural observations indicate that a phase transition from the solid to the liquid state has its onset between 8 and 20 GPa and that the liquid fraction is progressively increased with pressure. Which state is produced during adiabatic expansion depends on peak pressure and the specific material properties that are consolidated in the particular Hugoniot curves (Ahrens and O'Keefe, 1972). However, pure shock-heating was likely not sufficient to produce melt, as discussed above. Nevertheless, compression-induced shear movements along grain boundaries (Stöffler et al., 1991) and cracks additionally favor melting and may have played a significant role in the experiments of this study since melt exclusively occurs at grain rims or as single fragments (Fig. 2.4.8b - d), but never enclosed in a crystalline environment. Instead, the opposite occurs (Fig. 2.4.8c). The sharp contact between melt and the crystalline phase is not only indicative for rapid cooling, but shows that melting occurred as a clearly limited process. This observation indicates that melting is triggered by friction and shock heating, similar to as it occurs on a larger scale in pseudotachylites of impact craters (Reimold, 1998). Wheather friction, shock heating or a combination of both produces pseudotachylites in impact craters is still under debate, but similar veins can be exclusively formed by friction heating in shear zones (e.g. Passchier and Trouw, 2005). Regardsless of the exact mechanism, the contact between the amorphous and the crystalline phase is relatively sharp and the fact that melt in the 30 GPa sample occurs at the outer grain rims supports this hypothesis. As mentioned, grain rims likely display former cracks along which fracture collapse as well as friction producing movements are strongly favored when stress is imposed.

In contrast, the amorphous domains and rippled Fe-accumulations (Fig. 2.4.9c - d) are clearly shock-induced. Especially the rippled Fe-accumulations (Fig. 2.4.9c) suggest a pure shock-triggered process that leads to a change in physical state. In order to shed light onto the striking composition of these iron-rich sulfur phases, the phase relations in the Fe-FeS system will be briefly discussed in the following.

Reactions in this system are well-suited for comparison as they are similar to those of pyrrhotite (Anderson and Ahrens, 1986). Troilite and pure iron with eutectic-like textures can be simultaneously formed when a sulfur-rich iron is quenched (Brett and Bell, 1969). The produced iron is described by Brett and Bell (1969) as rosettes, stars and dendrites. These are textures that are identical with those from the quenched melt in the experiments of this study. Eutectic iron sulfide compositions are pressure-dependent and incorporate a larger iron content with increasing pressure, as shown by Usselmann (1975a + b). The studies of Usselmann (1975a) show that various compositions of iron sulfides can be formed upon cooling and heating, whereby the melting/crystallizing phase segregates from the present phase. Structure and composition of the iron sulfides in the 20 GPa sample indicate that such processes first happen at the atomic scale. Furthermore, the estimated temperatures in the container suggest that this effect was dominantly triggered by pressure rather than by temperature. It is known that decomposition of shock melt can be strongly triggered by pressure release (Fritz and Greshake, 2009). This process, however, has not been brought to completeness in the 20 GPa sample since the conditions that allowed these "plastic-liquid" deformations changed abruptly. Therefore, the transformation state was "frozen". This process is distinctly different to heat-melting, since changes in temperature manifest distinctly slower in the material than changes in pressure. Furthermore, these results show that shock melting is presumably accelerated by friction heating and the effect of shock, and first establishes in the material without affecting the optical properties at the μm scale. The pressure range within which those deformation features develop are different to those observed in troilite. Schmitt (2000) did not describe any evidence of melting in troilite when shocked at 293 K up to 60 GPa. On the other hand, Joreau et al. (1996) found two sets of PDFs in troilite, which was part of shocked chondrites that suffered shock pressures above 45 GPa. Melting of troilite has been observed above 45 GPa when the sample was pre-heated to 920 K (Schmitt, 2000). Melt droplets in troilite could be found by Benett and McSween (1996) in L-group ordinary chondrites, which suffered shock pressures above 20 GPa. Pre-shock temperatures of these samples are, however, unknown.

High pressure phase?

It has been repeatedly discussed weather pyrrhotite forms an hpp. Even though a phase transition has been observed under maintaining pressure at~6.8 GPa (Kamimura et al., 1992), the nature of this hpp remains unclear. Ahrens (1979) suggested this transition to occur between 2.7 and 3.8 GPa and concluded from calculated post-shock densities that this phase remains after shock. The same is observed from hpps from silicates that have been found in natural impact craters (Gillet et al., 2000; Langenhorst and Poirier, 2000; Sharp et al., 1997, 1999). On the other hand, Mössbauer spectra suggest that the transition is not a structural one, but is mainly restricted to a change in compressibility of the c-axis (Kobayashi et al., 1997). Magnetic properties of pyrrhotite shocked up to 12 GPa (Louzada et al., 2010) show no indication of a relictic hpp, which is also the case in the study performed for this thesis. The magnetic measurements illustrate that monoclinic 4C and hexagonal NC pyrrhotite are present in all samples and the lattice parameters (Table 2.4.3) clearly confirm these results. It can be therefore excluded that a crystalline hpp phase is present in the pyrrhotite samples shocked up to 30 GPa. On the basis of the proved transition under maintaining pressure (Gilder et al., 2011; Kamimura et al., 1992), this conclusion allows two different interpretations: First, the hpp is immediately transformed back after shock, or second, the hpp phase remains but is only sparsely present within the dominant lpp.

The first scenario is possible if the transition is not structural (Kobayashi et al., 1997) and restricted to elastic compression of the crystal parameters under pressure. In this case, the transition is not a "real" phase transition, but in fact displays a reversible change in magnetic ordering under pressure resulting in a paramagnetic state with quadrupole splitting (Kobayashi et al., 1997). An hpp as suggested by Ahrens (1979) for pyrrhotite, however, possesses a denser atomic ordering compared to their low-pressure equivalent. Such phases generally crystallize from a dense high-pressure liquid or polymineralic high-pressure melt during pressure release (Greshake et al., 2000) and display permanent state transitions. It is known from most of these phases that they remain stable after shock (e.g. Ahrens, 1979; Kleeman and Ahrens, 1973; Leroux, 2001; Stöffler et al., 1991), which is plausible since such extremely short-time scaled reactions do not provide sufficient thermal energy to activate a complete structural back-transformation.

The second scenario implies that the hpp phase initially forms at a low fraction which increases with shock pressure. This mechanism is suggested for pyrrhotite by Ahrens (1979) who estimated that the process is brought to completeness at~25 GPa, but this is very likely not the case, at least not in the experiments presented here. However, this process is of particular interest especially for the upper pressure regime between 20 and 30 GPa as well as for the amorphous domains that are formed therein. It is known from hpp's in silicates that such phases are often associated with glass fractions (Kieffer et al., 1976; Kitamura et al., 1977), also described as short-range order (SRO) phase (Kleemann and Ahrens, 1973). High-pressure melt that is not transformed into hpp appears to remain as a dense amorphous phase (Greshake et al., 2000) and could be the zero pressure equivalent to the dense high pressure SRO phase (Kleeman and Ahrens, 1973). For SiO_2, the transformation from a high-pressure crystal to an amorphous phase occurs in a thermodynamic equilibrium state at which the formation of a stable crystalline phase is prevented by kinetic effects (Williams and Jeanloz, 1989). Amorphous structures are also reported for troilite (Joreau, 1996), but here they are restricted to the framework of parallel lamellae (most likely PDFs) or form quenched melt. This leaves three possibilities to interpret the amorphous spots of this study: Either those domains are relicts of the hpp, have been formed as a precursor of shock melting, or simply are an equivalent to diaplectic glass. In either cases they developed during decompression (Kieffer et al., 1976; Kleeman and Ahrens, 1979; Williams and Jeanloz, 1989). Further experiments exceeding 30 GPa should be undertaken to shed light on this problem. If a hpp is observed in such experiments, this should weaken the ferrimagnetic character of the bulk phase (Kamimura et al., 1992; Kobayashi, 1997).

2.4.5 Conclusions

Rock magnetic properties of shocked pyrrhotite are controlled by complex interactions between different shock deformation features. In the low-pressure regime up to 8 GPa, mechanical deformation and associated microstructures dominate and tend to increase in frequency. As observed in earlier shock experiments on pyrrhotite (Louzada et al., 2007, 2011), the most obvious magnetic feature is an increase of the SD character. The microstructural observations of this study suggest that this effect is not only ascribed to grain size reduction, but also to a re-arrangement of magnetic domains. Amorphous lamellae determine the area within which magnetic domains can develop mostly by means of their frequency and the particular distance to each other. The 20 - 30 GPa pressure regime is dominated by rather complex diffusion processes. Rippled Fe-accumulations and amorphous spots can be found at 20 GPa and additional features are the formation of melt as well as abundant deformation twins at 30 GPa. The extreme short time-interval within which external conditions change when shock is applied and removed, hampers chemical equilibration and leads to "freezing" of uncompleted reactions and their associated microstructures. The best examples for this are rippled Fe-accumulations in the 20 GPa sample which vary in composition within few nm. Moreover, those structures reveal that shock-induced microstructures initially occur on a submicron scale and progressively develop with increasing pressure. The rippled Fe-accumulations in the 20 GPa sample represent a kind of "transition state" and can lead to confusion since iron is accumulated at the submicron scale, but the samples seems to be macroscopically free of structures that indicate melting or other chemical reactions. In such a state, one observes brittle deformation with optical or scanning electron microscopes, whereas nm-scale physical state changes already affect the crystal lattice and therewith the magnetic properties. Besides shock deformation features, the formation of secondary minerals is therefore an additional important feature that changes the rock magnetic properties at higher pressures. For pyrrhotite, native iron is probably the most important one. Prospecting of iron seems to be a good indicator for diffusion processes and can be relatively easy achieved by χ-T measurements. However, small particles of native iron quickly oxidize in air as it happened to the samples of this study after the first measurement campaign. The magnetic signal of pure iron is then no longer present in the obtained data. Microstructures and secondary phase formation can therefore imply crucial information about consequences of shock on pyrrhotite, but the extremely small average size make them also very instable, especially under natural conditions.

The results show that neither chemical composition nor crystal structure are modified if shock is applied on pyrrhotite. However, melting and the segregation of native iron do change the chemical composition, but not in a homogeneous way. Although the rippled Fe-accumulations are too small to be detected by the microprobe tool because the resolution of the instrument is too low, these structures should affect the measurements in some way. Segregated iron is easily oxidized under natural conditions and this process is usually associated with the diffusion of iron (Pratt, 1994). An oxidized grain containing such features may therefore become depleted in iron. Nevertheless, the diffusion of iron either produces a random or new atomic ordering in the affected areas. In the latter case, the shock-induced deformation features are erased by the reordering process. Melting and crystallization of native iron should not lead to confusion since this process is easily detectable by the microprobe instrument. Although reflections in the diffraction pattern are slightly widened, streaking is not present revealing that streaking is not a consequence of shock deformation. This indicates that the local density of lattice defects introduced by shock is too low to produce this feature in the diffraction pattern, at least in the crystalline parts of the post-shock grains.

Chapter 3

Final conclusions

Final conclusions

Magnetite and pyrrhotite are the most important magnetic minerals on Earth. Therefore, understanding the effect of shock waves on these magnetic minerals is critical for the interpretation of magnetic anomalies around impact structures. In this thesis both minerals have been thoroughly investigated with respect to their structural and magnetic properties which were modified by different processes linked with the development of the Chesapeake Bay Impact Structure (CBIS). Since the results of the naturally shocked samples revealed that alteration in such settings modifies some of the shock features, an experimental explosive setup was applied on pyrrhotite, in order to study the direct shock deformation features and their consequences on the magnetic properties of this mineral. In this last chapter the results of these experiments will be compared with the observations from the CBIS and their general conclusions will be discussed.

3.1 Iron-deficient and experimentally shocked pyrrhotite

The results from the shock experiments of this study revealed that several discrepancies exist between iron-deficient and experimentally shocked pyrrhotite. Iron-deficient pyrrhotite contains abundant stacking faults that were interpreted as an evidence of shock treatment. The frequency of stacking faults is so high that reflections in the SAED pattern are largely streaked (2.3.4a). This streaking makes it impossible to determine the superstructure of that mineral, but ferrimagnetism occurring below a Curie temperature (T_C) of 345 - 360°C reveals that an ordered structure is, at least partially, present. Enhanced T_Cs and a missing 34 K transition, in return, show that iron-deficient pyrrhotite is different to 4C pyrrhotite. The iron-deficient composition is homogeneously distributed within a single grain, but variations of ~1% occur between all grains of a sample concerning the percentage of iron. In general, iron-deficient pyrrhotite lies within the composition range of smythite approximately fitting the formula Fe_9S_{11} (Fig. 2.3.9). It is furthermore characterized by the main reflections of the NiAs cell, typical for all pyrrhotite modifications. Missing transition temperatures, typical for 4C pyrrhotite, as well as the homogeneous composition indicate that no relict pyrrhotite is present in the studied samples. Therefore, transformation from pyrrhotite to iron-deficient pyrrhotite must have occurred pervasively and this process has thus produced a new ferrimagnetic ordering. Since the abundant stacking faults, however, do distort the crystal structure, the size of the magnetically ordered domains is suggested to lie in the range of the average distance between two adjacent stacking faults, which is ~10 nm.

The shock experiments, however, revealed that shock pressures between 3 and 30 GPa alone do not change the chemical composition of pyrrhotite. In fact, various shock deformation features develop. In the lower pressure regime up to 8 GPa, brecciation and the formation of stacking faults as well as lamellar amorphous domains are predominant. Between 20 and 30 GPa amorphous domains occur as patches in the crystal lattice, and those domains increase in size with pressure (Fig. 2.4.9). However, grain size reduction has the largest effect on the rock magnetic properties. The latter are generally characterized by a magnetic hardening and suppression of the magnetic transitions in the lower pressure regime. The importance of these effects then again decreases in the upper pressure regime. Findings from the shock experiments therefore reveal that the iron-deficiency and the establishment of a new ferrimagnetic structure in iron-deficient pyrrhotite are both not shock-induced, but more likely the result of another process. Since no evidence was found in the natural samples indicating that iron-deficient pyrrhotite crystallized from impact melt, the most likely driving mechanism of these processes is low-temperature (LT) oxidation. In natural environments LT oxidation is the usual process producing smythite from 4C pyrrhotite (e.g. Jover et al., 1989; Taylor and Williams, 1972). While iron diffuses out of the crystal (Fleet, 1982) during this transformation, associated vacancy movements cause partial reordering of the atomic structure (Izola et al., 2007; Van Landuyt and Amelinckx, 1972).

Since one important conclusion of this study is that iron-deficient pyrrhotite with the smythite-like structure is related to LT oxidation, the question remains weather the transformation of this phase occurred before or after the impact event. The crucial point concerning this question is probably the origin of the lattice defects in iron-deficient pyrrhotite. Assuming that the high defect concentration is a product of shock deformation, post-impact alteration is hard to bring in line with the abundant defects in iron-deficient pyrrhotite. If post-impact alteration formed the smythite structure, one would expect that the defects were largely annealed by the reordering process. A logical interpretation is therefore the formation of a smythite-like phase during basement alteration occurring prior to the impact event. Additional to the lattice defects, which are described to occur during the transformation from pyrrhotite to smythite (Fleet, 1982), shock-induced defects where then superimposed onto the crystal structure. Since iron-deficient pyrrhotite has already been present when the impact occurred, the composition did not change with the shock. This observation is in agreement with the observations of the experimental part of this study. In this case, both defect generations are impossible to be separated by the methods applied in this study. By all means, post-impact alteration affected the suevite and transformed parts of the iron-deficient pyrrhotite grains into marcasite/pyrite (Fig. 2.3.2). However, no smythite-containing host rock was found in this study. Although such rock units have not been encountered by the four sampled crater drillings, it is most likely that iron-deficient pyrrhotite derives from the schist rock. This unit is the only rock in which pyrrhotite was found. If it was affected by pervasive pre-impact alteration, both anomalous and iron-deficient pyrrhotite were likely formed during that event. In this case, alteration intensity was not homogeneously distributed within the schist. Apart from the formation of secondary marcasite/pyrite, post-impact alteration has then not produced significant changes in chemistry of the pre-altered phases.

This interpretation, however, leaves open some important questions. It has been shown that post-impact alteration strongly affected grains of secondary magnetite and associated chemical processes furthermore produced secondary minerals as magnetite or chlorite and calcite. Why are the iron-deficient pyrrhotite grains largely unaffected by this process so that only small parts of such grains suffered post-impact transformations into marcasite/pyrite? Furthermore, why do, besides the stacking faults, none of the deformation structures observed in experimentally shocked pyrrhotite appear in the natural samples? In this context it has to be considered that shock pressures of 10 - 35 GPa have been postulated for the CBIS (Wittmann et al., 2009a), and the shock experiments revealed that striking deformation structures like PDFs already form at 8 GPa. Although the latter were not yet observed at 20 GPa, other features, as for example amorphous domains, appear in this pressure range, but none of those features appear in the natural samples. It must therefore be questioned if iron-deficient pyrrhotite was really formed before the impact event. If it was not, the abundant stacking faults were consequently not a product of shock. An important hint on this topic is probably given by the work of Fleet (1982) who produced synthetic smythite from pyrrhotite. In his experiments Fleet (1982) could show that the mentioned transformation produces such a high concentration of stacking faults in the crystal lattice that diffraction streaks occur. Analogous to iron-deficient pyrrhotite, streaking is observable in the inner row of the diffraction pattern (Fig. 2.3.2b, 2.3.4a and Fig. 3 in Fleet, 1982). Interestingly, such streaking was not observed in the diffraction pattern of experimentally shocked pyrrhotite. The only visible influence of crystal defects on the SAED pattern of these samples is a slight reflection broadening (Fig. 2.4.9e). This leads to the conclusion that the density of stacking faults is distinctly lower in the experimentally shocked samples, since streaking is described to be a result of abundant stacking faults (Fleet, 1982; Mang et al., 2011). In other words, defect-free domains in experimentally shocked pyrrhotite must be distinctly larger than in iron-deficient pyrrhotite or smythite.

Based on these considerations, no valid argument remains stating that the stacking faults in iron-deficient pyrrhotite are the result of shock. From this point of view it seems rather likely that pyrrhotite has first been shocked and pervasively been oxidized afterwards. The reason why the shock-deformation features observed in experimentally shocked pyrrhotite are missing in the CBIS samples, is therefore most likely the discussed transformation process which has erased large parts of such structures. This conclusion is confirmed by three more observations. First, shocked pyrrhotite from the Bosumtwi impact crater, Ghana, has probably been equally transformed into smythite, but these grains still contain larger fractions of pyrrhotite (Kontny et al., 2007). HRTEM photographs of such grains show defect structures which are similar to the PDFs of the 8 GPa sample from this study (Fig. 2.4.9a and Fig. 2 in Kontny et al., 2007). As mentioned before, such structures do not appear in iron-deficient pyrrhotite. Secondly, the homogeneous composition of iron-deficient pyrrhotite grains (Fig. 2.3.6) argues in favour of a pervasive process which did not spare appreciable remnants of 4C pyrrhotite and associated deformation structures. In contrast, some of the pyrrhotite grains of Bosumtwi show a clear rim of 4C pyrrhotite/smythite surrounding a core of antiferromagnetic NC pyrrhotite (Fig. 1e in Kontny et al., 2007). The latter represents a more iron-rich member of the pyrrhotite system and is often the unoxidized precursor product of 4C pyrrhotite (Kontny et al., 2000). Thirdly, the twin domains of iron-deficient pyrrhotite look entirely different to those observed in the 30 GPa sample from the shock experiments (Fig. 2.3.1d and 2.4.9e + f, respectively). The former have an average width of 10 nm and the twin boundaries, wich probably arise from the reordering process, are not well-defined. In contrast, the deformation twin domains in experimentally shocked pyrrhotite have an average width of 50 nm and show clearly defined boundaries.

Taking everything into account, it seems more likely that iron-deficient pyrrhotite is a product of post-impact rather than of pre-impact alteration. If so, the formation of anomalous pyrrhotite in the schist and suevite is equally more a result of post-impact oxidation, which is a more realistic explanation. A crucial process for both anomalous pyrrhotite and smythite is obviously grain size reduction, which largely increases the specific grain surface. Since pyrrhotite from the schist occurs in the host rock, it is plausible that the oxidation process is less pervasively developed in these grains than in the distinctly smaller pyrrhotite grains from the suevite. Deposition of the suevite sediments was associated with a large incorporation of sea water (Horton et al., 2005a), weakening the rock strength. Fluid mobility was therefore distinctly larger in this unit.

Nonetheless, the scenario explaining the origin of iron-deficient pyrrhotite by post-impact alteration is largely based on the findings of Fleet (1982). Unfortunately smythite remains rather poorly investigated up to now, and further work studying the transformation process from pyrrhotite to smythite do not yet exist to the knowledge of the author. On the one hand, the results of this thesis give therefore new important insights concerning the relation between smythite and pyrrhotite. On the other hand, future work on this topic is necessary to confirm the postulated conclusions of this thesis. Such should imply a study of smythite samples under the transmissions electron microscope (TEM) in order to verify the effect of reflection streaking.

3.2 Comparison of shock-related features between magnetite and pyrrhotite

Some of the features observed in shocked pyrrhotite are comparable with those from shocked magnetite. Such common features indicate a more general process affecting shocked magnetic minerals in impact structures. This study reveals two significant processes influencing the magnetic properties of impactites: grain size reduction and alteration of the magnetic minerals. Grain size reduction is largely triggered by brittle deformation, which is a pervasive and prominent feature associated with release of the shock wave (e.g. Melosh, 1989). This process creates a larger fraction of single domain (SD) grains, and the post-shock samples are generally characterized by increased coercivities, saturation isothermal remanent magnetization (SIRM) and LT memory, compared to the pre-shock samples. These results are in good agreement with the observations of former studies (Louzada et al., 2007, 2011; Gilder et al., 2002, 2004). However, this thesis reveals that deformation features on the submicron scale additionally decrease magnetic domain sizes. Planar fractures (PF), planar deformation features (PDF), and even amorphous domains are able to produce magnetic domains with SD-like behaviour in grains with sizes suggesting clear multidomain (MD) behaviour. Whereas PFs and PDFs subdivide the grains into small structural domains, amorphous spots largely reduce the crystalline parts in a shocked grain. If sufficiently developed, these spots produce a pattern consisting of amorphous and crystalline domains. Even though PDFs have been only observed in experimentally shocked pyrrhotite (Fig. 2.4.9a + b), PFs appear in naturally shocked magnetite as well (Fig. 2.2.2c). The average distance of these PFs is sufficient to produce a fraction of SD domains.

Fractures and cracks commonly facilitate post-impact alteration processes. It is generally accepted that alteration largely affects the impact units and that such processes are enhanced due to the increased surface of the smaller grains (e.g. Wittmann et al., 2009a). The results of this study suggest that crystal defects additionally favour such diffusion processes. Iron-deficient pyrrhotite, for example, has been strongly oxidized, but reaction rims indicating the diffusion of Fe out of the particular grains are missing. Therefore it seems likely that internal defects strongly favoured diffusion and allowed this process to develop pervasively. Alteration can also modify the rock magnetic properties, but here different processes may contribute to the bulk rock magnetic properties. For example, the domains in iron-deficient pyrrhotite, which have been transformed to marcasite/pyrite, have lost their ferrimagnetic properties as a consequence of the modification process. This leads to a lowering of the bulk rock magnetic properties of the suevite. In contrast, the transformation from goethite to magnetite has raised the rock magnetic properties of this unit. Whereas pyrrhotite has been entirely transformed into new phases with different magnetic properties, those of shocked magnetite have mostly been retained. However, slight changes in chemical composition lead to a partial non-stoichiometry and do modify some important parameters of magnetite. Probably the most important one is T_V. In any case, lowering and depression of T_V is a result of oxidation and has no direct relation to shock. However, shock does have an indirect influence since it is responsible for grain sizes reduction. The thereby increased surface is an important pre-condition for the alteration process. Naturally shocked magnetite and pyrrhotite from this study show that post-impact alteration can strongly modify the particular shock features. Consequently, conservation of these features is largely dependent on the post-impact alteration history.

3.3 Natural remanent magnetization

The rock units from the CBIS are commonly characterized by an unstable natural remanent magnetization (NRM) vector (chapter 2.1). This is in contradiction to observations from other impact structures which often show the dominance of a stable magnetization direction acquired in the magnetic field at the time of the impact (e.g. Bosumtwi: Kontny et al., 2007, Ries: Pohl et al., 2010). Only few suevite and schist samples from the CBIS show stable directions and these directions can be directly linked to secondary magnetite and iron-deficient pyrrhotite in the suevite, and anomalous pyrrhotite in the schist. If iron-deficient pyrrhotite is a product of post-shock alteration, this mineral clearly carries a chemical remanent magnetization. This is confirmed by some samples containing both secondary magnetite and iron-deficient pyrrhotite. These samples show only one stable direction during demagnetization indicating that this direction is stable in both minerals. A CRM is usually acquired below the blocking temperature when a superparamagnetic (SP) crystal enters the SD size range during crystal growth (Özdemir and Dunlop, 1993). This process is associated with large changes in magnetic relaxation times hampering a spontaneous flip of the remanence vector (Haigh, 1957; Kobayashi, 1959; Nagata, 1971). In Return, the remarkable fraction of SP grains in secondary magnetite is likely dominating the induced magnetization which is present in few suevite samples (Dunlop and Özdemir, 1997).

In contrast, shocked magnetite does not show a stable magnetization, although some grains are in the SD range. If these grains acquired a shock remanent magnetization (SRM), this magnetization was not stable and easily destroyed by alteration. Shock experiments on natural samples by Gattacceca et al. (2007) revealed that the post-shock NRM is dependent on the coercivity of the initial magnetic carriers. Since these are supposed to form MD magnetite in the CBIS, the capacity to acquire a remanent magnetization should be drastically increased for the shocked samples. This becomes obvious comparing the rock magnetic properties of the granite megablock with those of the suevite and impact breccias. However, Gattacceca et al. (2007) obtained a large fraction of post shock samples without a stable direction. Within the fraction that carried a stable remanent magnetization the direction was not consistent. These results are similar to those of this study, and the question arises if the acquisition of an SRM may be linked to a certain temperature. Kontny et al. (2007) indeed explained the stable directions of shocked pyrrhotite in the Bosumtwi impact crater by elevated temperatures, which were present after the impact. These authors argued that such temperatures were close to the blocking temperature (T_B) of pyrrhotite, which is why such grains could acquire a stable magnetization. During formation of the CBIS, however, incorporation of large amounts of seawater caused rapid cooling of the crater temperatures (Wittmann et al., 2009a). Alteration temperatures were probably distinctly lower than the T_B of pyrrhotite in the CBIS and no stable magnetization was acquired in those grains. Furthermore, for the acquisition of a stable direction the specific grain size is an important factor, since MD grains are generally not able to record a specific magnetization. This happens because magnetic domain walls in these grains can be relatively easily shifted when the external field changes (Dunlop and Özdemir, 1997). Compiling all these results, it seems likely that an SRM is hard to acquire when the temperatures remain distinctly below the blocking temperature of a magnetic mineral.

3.4 General relevance for magnetic properties of impact structures

Impact craters are prominent features which modify large parts of the Earth surface and even larger parts of extraterrestrial bodies like Moon, Mars or asteroids (Ivanov, 2001; Reimold and Koeberl, 2008; Stöffler et al., 2006). Magnetic anomalies associated with such structures are often complex, but contain important information about target rocks, shock wave intensity, and post-shock alteration. Basically, the magnetization of rocks is, in return, profoundly dependent on the presence or absence of the most important magnetic carriers. In general, these carriers are magnetite, pyrrhotite, and hematite (Fe_2O_3) and less frequently smythite (Fe_9S_{11}) and greigite (Fe_3S_4). Less important are goethite (FeOOH) and siderite ($FeCO_3$), since their contribution to the bulk magnetization is usually low. In general, shock can result in partial demagnetization of rocks. This mechanism is evidenced by large impact basins on Mars (Rochette et al., 2003, 2005) and experimental studies (e.g. Louzada et al., 2007). Therefore the bulk magnetization of a particular rock is additionally dependent on the physical state of the particular magnetic carriers. The results of this study show that the most important factors controlling the particular magnetic properties are grain size, alteration state, and atomic ordering. However, the bulk magnetic field measured over an impact structure like the CBIS is also dependent on larger-scaled processes. In this connection, the most important ones are: randomization of the impact units during re-deposition, de- and remagnetization during and after shock, the formation of new magnetic minerals from impact melt or during hydrothermal activity, and the formation of new magnetic phases in the post-impact sediments (Hart et al, 1995; Pilkington and Grieve, 1992; Pohl et al., 2010). For modelling of the subsurface on the basis of the surface magnetic anomaly pattern, usually a constant magnetization is attributed to a specific rock. A structural model is then developed with the help of information obtained from seismic surveys or drillings. A quite reasonable model for the CBIS has been proposed by Shah et al. (2009). In the study of these authors the magnetic highs are explained by the presence of basement megablocks, whereas the suevite unit is considered as a uniformly magnetized layer. The results from this study reveal though, that this assumption is largely simplified, since the suevite layer is extremely inhomogeneous in composition. This unit contains secondary and shocked magnetite as well as iron-deficient pyrrhotite, and the appearance of each is locally restricted. This observation indicates that all these minerals occur in close distance to their potential source rocks. Secondary magnetite, for example, occurs much more abundantly close to impact melt fragments whose decomposition probably favoured the formation of the initial iron-oxide. Especially in the suevite the inhomogeneous spatial distribution of magnetic minerals can create magnetic anomaly patterns that are not correctly interpreted if a uniform magnetization is assumed. Areas showing a high frequency of magnetite or pyrrhotite can significantly contribute to a short-wave magnetic anomaly pattern, but may be mistaken with basement blocks. Hence, the study of Shah et al. (2009) has to be complemented by the fact that not only blocks, megablocks, or melt sheets, but also local parts of the suevite unit are responsible for the short-wavelength magnetic anomaly peaks.

In impact craters, melting is another important and general shock-related feature. Although no magnetic minerals crystallized from the impact melt, fragments have been observed in this study. This process is of great importance in other impact craters (Pilkington and Grieve, 1992; Pohl et al., 2010). The formation of new magnetic minerals carrying a TRM should be of larger importance regarding compact sheets of impact melt and pseudotachylites, since the formation of small magnetic phases in the SD range from a melt phase is largely known in other settings like oceanic basalts (Gee and Kent, 1997). The results of this thesis show that the presence of impact melt is additionally important for the formation of secondary magnetic minerals. Additionally, shock-melting of magnetic minerals can form new magnetic phases with different magnetic properties. In the shock experiments of this study, for example, native iron formed as a new mineral from molten pyrrhotite. If this process affects a larger fraction of the magnetic carriers, it has a remarkable effect on the bulk rock magnetic properties of a particular unit.

In many terrestrial craters, magnetic anomaly pattern associated with the crater structure are often restricted to the inner crater rims. For example, this is true for the Ries crater, Germany (Pohl et al., 2010) or the Bosumtwi crater, Ghana (Ugalde et al., 2007). But since the impact at Chesapeake Bay occurred on a continental shelf margin, sea water was involved in the crater modification stage. The strong ocean resurge was responsible for large modifications of the crater structure occuring beyond the inner crater. Such modifications included the displacement of rock blocks and enhanced alteration of impact sediments and basement units (Ormo and Lindström, 2000; Powars and Bruce, 1999). This could explain why the magnetic anomaly pattern is not restricted to the inner crater rim in the CBIS. As a result, the formation of the crater structure is dependent on the initial geological setting of the particular impact location. The geological setting is therefore an additional factor influencing the development of the associated magnetic anomaly pattern.

References

Abadian, M., 1972: Petrographie, Stoßwellenmetamorphose und Entstehung polimikter kristalliner Breckzien im Nördlinger Ries. *Contributions to Mineralogy and Petrology* (35), 245-262.

Ahrens, T. J., 1979: Equations of state of iron sulfide and constraints on the sulfur content of the Earth. *Journal of Geophysical Research* (84), B3, 985-998.

Ahrens, T.J., O'Keefe, J.D., 1972: Shock melting and vaporization of lunar rocks and minerals. *Earth, Moon and Planets* (4), 1-2, 214-249.

Anderson, W.W., Ahrens, T.J., 1986: Shock wave experiments on iron sulfide and sulfur in planetary cores, *Lunar and Planetary Science (XVII)*, 11-12.

Aragón, R., Buttrey, D.J., Shepherd, J.P., Honig, J.M., 1985: Influence of nonstoichiometry on the Verwey transition. *Physical Review B* (31), 1, 430-436.

Archanjo, C.J., Laune. P, Bouchez, J.L., 1995: Magnetic fabric vs. magnetite and biotite shape fabrics of the magnetite-bearing granite pluton of Gameleiras (Northeast Brazil). *Physics of the Earth and Planetary Interiors* (89), 63-75.

Arnold, R.G., 1962: Equilibrium relations between pyrrhotite and pyrite from 325 degrees to 743 degrees C. *Economic Geology* (57), 1, 72-90.

Artemieva N. A., Ivanov B., 2004: Launch of Martian meteorites in oblique impacts. *Icarus* (171), 84-101.

Ashworth, J.R., Schneider, H., 1985: Deformation and transformation in experimentally shock-loaded quartz. *Physics of Chemistry and Minerals* (11), 241-249.

Bartosova, K., Ferrière, L., Koeberl, C., Reimold, W.U., Gier, S., 2009: Petrographic and shock metamorphic studies of the impact breccia section (1397-1551 m depth) of the Eyreville drill core, Chesapeake Bay impact structure, USA. *Geological Society of America Special Paper* (458), 317-348.

Bennett, M.E., McSween, H.Y., Jr., 1996: Shock features in iron-nickel metal and troilite of L-group ordinary chondrites. *Meteoritics and Planetary Science* (31), 255-264.

Bertraut, P.E.F., 1953: Contribution à l'étude des structures lacunaires: La Pyrrhotine. *Acta Crystallographica* (6), 557-561.

Bezaeva, N. S., Gattacceca J., Rochette P., Sadykov, R.A. and Trukhin V.I., 2010: Demagnetization of terrestrial and extraterrestrial rocks under hydrostatic pressure up to 1.2 GPa. *Physics of the Earth and Planetary Interiors* (179), 7-20.

Brett, R., Bell, P.M., 1969: Melting relations in the Fe-rich portion of the system Fe-FeS at 30 kb pressure. *Earth and Planetary Science Letters* (6), 479-482.

Butler, R.F., 1998: Paleomagnetism: Magnetic domains to geologic terranes. Electronic edition, www.geo.arizona.edu/Paleomag/book/.

Cao, L.-F., Xie, D., Guo, M.-X., Park, H.S., Fujita, T., 2007: Size and shape effects on Curie temperature of ferromagnetic nanoparticles. *Transactions of Nonferrous Metals Society of China* (17), 1451-1455.

Carporzen, L., Gilder, S.A., Hart, R.J., 2006: Origin and implications of two Verwey transitions in the basement rocks of the Vredefort meteorite crater, South Africa. *Earth and Planetary Science Letters* (251), 305-317.

Carter-Stiglitz, B., Moskowitz, B., Solheid, P., Berquó, T.S., Jackson, M., Kosterov, A., 2006: Low-temperature behavior of multidomain titanomagnetites: TM0, TM16 and TM35. *Journal of Geophysical Research* (111), B12S05, doi: 10.1029/2006JB004561.

Catchings, R.D., Powars, D.S., Gohn, G.S., Horton, J.W., Jr., Goldman, M.R., Hole, J.A., 2008: Anatomy of the Chesapeake Bay impact structure revealed by seismic imaging, Delmarva Peinisula, Virginia, USA. *Journal of Geophysical Research* (113), B08413,
doi: 10.1029/2007JB005421.

Chen, A.P., Egli, R., Moskowitz, B.M., 2007: First-order reversal curve (FORC) diagrams of natural and cultured biogenetic magnetic particles. *Journal of Geophysical Research* (112), B08S90, doi: 10.1029/2006JB004575.

Clark, A.H., 1966: Stability field of monoclinic pyrrhotite. *Transactions of the Institution of Mining and Metallurgy* (75), B232-B235.

Clark, J.F., 1983: Magnetic survey data at meteoritic impact sites in North America. *Geomagnetic Survey Canada Earth and Physics Branch, Branch* Open File Report (83-5), 1-30.

Collins G.S., Wünnemann K., 2005: How big was the Chesapeake Bay impact? Insight from numerical modeling. *Geology* (33), 12, 925-928.

Day, R., Fuller, M., Schmidt, V.A., 1977: Hysteresis properties of titanomagnetites: Grain-size and compositional dependence. *Physics of the Earth and Planetary Interiors* (13), 260-267.

Dekkers, M.J., 1990: Magnetic monitoring of pyrrhotite alteration during thermal demagnetization. *Geophysical Research Letters* (17), 6, 779-782.

Dekkers, M.J., Mattéi, J.-L., Fillion, G., Rochette, P., 1989: Grain-size dependence of the magnetic behavior of pyrrhotite during its low- temperature transition at 34 K. *Geophysical Research Letters* (16), 8, 855-858.

Desborough, G.A., Carpenter, R.H., 1965: Phase relations of pyrrhotite. *Economic Geology* (60), 7, 1431-1450.

Deutsch, A., Koeberl, C., 2006: Establishing the link between the Chesapeake Bay impact structure and the North American tektite strewn field: The Sr-Nd isotopic evidence. *Meteoritics and Planetary Science Letters* (41), 5, 689-703.

Dódony, I., Pósfai, M., 1990: Pyrrhotite superstructures, part II: A TEM study of 4C and 5C structures. *European Journal of Mineralogy* (2), 529-535.

Dunlop, D.J., 1995: Magnetism in rocks. *Journal of Geophysical Research* (100), B2, 2161-2174.

Dunlop, D.J., 2002: Theory and application of the Day Plot (M_{rs}/M_s versus H_{cr}/H_c) 1. Theoretical curves and tests using titanomagnetite data. *Journal of Geophysical Research* (107), B3, 2056, doi: 10.1029/2001JB000486.

Dunlop, J.D., Özdemir, Ö., 1997: Rock Magnetism. *Cambridge University Press*, Cambridge, 573 p.

Duvall, G. E., Fowles, G. R., 1963: Shock waves. *In*: High pressure physics and chemistry 2, edited by Bradley R.S., *Academic Press Inc.*, London, New York, 209-291.

Edwards, L.E., Powars, D.S., 2003: Impact damage to dinocysts from the late Eocene Chesapeake Bay event: *Palaios* (18), 3, 275-285.

El Goresy, A. 1964: Die Erzmineralien in den Ries und Bosumtwi-Krater-Gläsern und ihre genetische Deutung. *Geochimica et Cosmochimica Acta* (28), 1881-1884.

Elbra, T., Kontny, A., Pesonen, L. J., 2009: Rock-magnetic properties of the ICDP-USGS Eyreville core, Chesapeake Bay

References

impact structure, USA. *Geological Sociecty of America Special Paper* (458), 119-136.

Feldmann, V.I., Sazonova, L.V., Milyavskii, V.V., Borodina, T.I., Sokolov, S.N., Zhunk, A.Z., 2006: Shock metamorphism of some rock-forming minerals. *Physics of the Solid Earth* (42), 6, 477-480.

Fleet, M.E., 1978: The pyrrhotite-marcasite transformation. *Canadian Mineralogist* (16), 31-35.

Fleet, M.E., 1982: Synthetic smythite and monoclinic Fe_3S_4. *Physics and Chemistry of Minerals* (8), 241-246.

French, B.M., 1998: Traces of catastrophe. A handbook of shock-metamorphic effects in terrestrial meteorite impact structures. *Lunar Planetary Institute Contributions* (954), Houston, 120 p.

Fritz J., Greshake A., 2009: High pressure phases in an ultramafic rock from Mars. *Earth and Planetary Science Letters* (288), 619-623.

Fritz J., Artemieva N. A., Greshake A., 2005: Ejection of Martian meteorites. *Meteoritics and Planetary Science* (40), 1393-1411.

Fritz J., Wünnemann K., Reimold W. U., Meyer C., Hornemann U., 2011: Shock experiments on quartz targets pre-cooled to 77 K. *International Journal of Impact Engineering* (38), 440-445.

Furukawa, Y., Barnes, H.L., 1996: Reactions forming smythite, Fe_9S_{11}. *Geochimica et Cosmochimica Acta* (60), 3581-3591.

Garrick-Bethell, J., Weiss, B., 2010: Kamacite blocking temperatures and applications on lunar magnetism. *Earth and Planetary Science Letters* (294), 1-7.

Gattacceca, J., Lamali, A., Rochette, P., Boustie, M., Berthe, L., 2007: The effects of explosive-driven shocks on the natural remanent magnetization and the magnetic properties of rocks. *Physics of the Earth and Planetary Interiors* (162), 85-98.

Gee, J., Kent, D.V., 1997: Calibration of magnetic granulometric trends in oceanic basalts. *Earth and Planetary Science Letters* (170), 377-390.

Gibson, R.L., Townsend, N.G., Horton, J.W., Jr., Reimold, W.U., 2009: Pre-impact tectonothermal evolution of the crystalline basement-derived rocks in the ICDP-USGS Eyreville B core, Chesapeake Bay impact structure. *Geological Society of America Special Paper* (458), 235-255.

Gilder, S.A., LeGeoff, M., Peyronneau, J., Chervin, J.-C., 2002: Novel high pressure magnetic measurements with application to magnetite. *Geophysical Research Letters* (29), 10, doi: 10.1029/2001GL014227.

Gilder, S.A., LeGoff, M., Chervin, J. C., Peyronneau, J., 2004: Magnetic properties of single and multi-domain magnetite under pressures from 0 to 6 GPa. *Geophysical Research Letters* (31), L10612, doi: 10.1029/2004GL019844.

Gilder, S.A., Egli, R., Hochleitner, R., Roud, S.C., Volk, M.W.R., Le Goff, M., de Wit, M., 2011: Anatomy of a pressure-induced, ferromagnetic-to-paramagnetic transition in pyrrhotite: Implications for the formation pressure of diamonds. *Journal of Geophysical Research* (116), B10101, doi: 10.1029/2011JB008292.

Gillet, P., Chen, M., Dubrovinsky, L., El Goresy, A., 2000: Natural $NaAlSiO_3O_8$-hollandite in the shocked Sixiangkou meteorite. *Science* (287), 1633-1636.

Gohn, G.S., Sanford, E.E., Powars, D.S., Hortin, J.W., Jr., Edwards, L.E., Morin, R.H., Self-Trail, J.M., 2004: Site report for the USGS test holes drilled at Cape Charles Northampton County, Virginia, in 2004. *U.S. Geological Survey* Open-File Report.

References

Gohn, G.S., Koeberl, C., Miller, K.G., Reimold, W.U., Cockell, C.S., Horton, J.W., Jr., Sanford, W.E., Voytek, M.A., 2006: Chesapeake Bay impact structure drilled. *Eos* (87), 35, 349.

Gohn, G.S., Koeberl, C., Miller, K.G., Reimold, W.U., Browning, J.V., Cockell, C.S., Horton, J.W, Jr., Kenkmann T., Kulpecz A. A., Powars D. S., Sanford W. E., Voytek M. A., 2008: Deep drilling into the Chesapeake Bay impact structure. *Science* (320), 1740-1745.

Gohn, G.S., Koeberl, C., Miller, K.G., Reimold, W.U. (editors), 2009a: The ICDP-USGS Deep Drilling Project in the Chesapeake Bay impact structure: Results from the Eyreville core holes. *Geological Society of America Special Paper* (458), 975 p.

Gohn, S.G., Koeberl, C., Miller, K.G., Reimold, W.U., 2009b: Deep drilling in the Chesapeake Bay impact structure-An overview. *Geological Society of America Special Paper* (458), 1-20.

Goltrant, O., Cordier, P., Doukhan, J.-C., 1991: Planar deformation features in shocked quartz; a transmission electron microscopy investigation. *Earth and Planetary Science Letters* (106), 103-115.

Goltrant, O., Leroux, H., Doukhan, J.-C., Cordier, P., 1992: Formation mechanisms of planar deformation features in naturally shocked quartz. *Physics of the Earth and Planetary Interiors* (74), 219-240.

Gose, W.A., Pearce, G.W., Strangway, D.W., larson, E.E., 1972: Magnetic properties of Apollo 14 breccias and their correlation with metanorphism. Geochimica et Cosmochimica Acta (3), 2387-2395.

Goya, G.F., Berquó, T.S., Fonseca, F.C., 2003: Static and dynamic magnetic properties of spherical magnetite nanoparticles. *Journal of Applied Physics* (94), 5, 3520-3528.

Grieve, R.A.F., Pesonen, L.J., 1992: The terrestrial impact cratering record. *Tectonophysics* (216), 1-30.

Greshake, A., Lingemann, C.M., Schmitt, R.T., Kenkmann, T., Stöffler, D., 2000: Pressure-temperature-time conditions for shock-produced high-pressure phases. *Meteoritics and Planetary Science* (35), supplement A65.

Grieve, R.A.F., Coderre, J.M., Robertson, P.B., Alexopoulos, J., 1990: Microscopic planar deformation features in quartz of the Vredeford structure: Anomalous but still suggestive of an impact origin. *Tectonophysics* (171), 185-200.

Grieve, R.A.F., Langenhorst, F., Stöffler, D., 1996: Shock metamorphism of quartz in nature and experiment: II. Significance in geosciences. *Meteoritics and Planetary Science* (31), 6-35.

Grønvold, F., Haraldsen, H., 1952: On the phase relations of synthetic and natural pyrrhotites ($Fe_{1-x}S$). *Acta Chemica Scandinavica* (6) 1452-1469.

Haigh, G., 1957: The process of magnetization by chemical change. *Philosophical Magazine* (3), 267-286.

Halgedahl, S.L., Jarrard, R.D., 1995: Low-temperature behavior of single-domain through multidomain magnetite. *Earth and Planetary Science Letters* (130), 127-139.

Haraldsen, H. 1937: Magnetochemische Untersuchungen, XXIV. Eine thermomagnetische Untersuchung der Umwandlungen im Troilit-Pyrrhotin-Gebiet des Eisen-Schwefel-Systems. *Zeitschrift für anorganische und allgemeine Chemie* (231), 78-96.

Harries, D., Pollok, K., Langenhorst, L., 2011: Translation interface modulation in NC-pyrrhotites: Direct imaging by TEM and a model towards understanding partially disordered structural states. *American Mineralogist* (96), 716-731.

Hart, R.J., Hargraves, R.B., Andreoli, M.A.G., Tredoux, M., Doucouré, C.M., 1995: Magnetic anomaly near the center of the Vredefort structure: Implications for impact-related magnetic signatures. *Geology* (23), 277-280.

References

Hecht, L., Wittmann, A., Schmitt, R.-T., Stöffler, D., 2004: Composition of impact melt particles and the effects of post-impact alteration in suevitic rocks at the Yaxcopoil-1 drill core, Chicxulub crater, Mexico. *Meteoritics and Planetary Science* (39), 7, 1169-1186.

References

Hedley, I.G., 1968: Chemical remanent magnetization of the FeOOH, Fe_2O_3 system. *Physics of the Earth and Planetary Interiors* (1), 103-121.

Heinrichs, D., 1967: Paleomagnetism of the Plio-Pleistocene Lousetown Formation, Virginia City, Nevada. *Journal of Geophysical Research* (72), 3, 277-294.

Henkel, H., 1992: Geophysical aspects of meteorite impact craters in eroded shield environment, with special emphasis on electric resistivity. *Tectonophysics* (216), 63-89.

Henkel, H., Reimold, W.U., 2002: Magnetic model of the central uplift of the Vredefort impact structure, South Africa. *Journal of Applied Geophysics* (51), 43-62.

Henkel, H., Reimold, W.U., Koeberl, C., 2002: Magnetic and gravity model of the Morokweng impact structure. *Journal of Applied Geophysics* (49), 129-147.

Hodych, J.P., 1990: Magnetic hysteresis as a function of low temperature in rocks: evidence for internal stress control of remanence in multi-domain and pseudo-single-domain magnetite. *Physics of the Earth and Planetary Interiors* (64), 21-36.

Hoffmann, V., 1993: Mineralogical, magnetic and Mössbauer data of smythite. *Studia Geophysica et Geadaetica* (37), 366-381.

Honig, J.M., 1995: Analysis of the Verwey transition in magnetite. *Journal of Alloys and Compounds* (229), 24-39.

Horton, J.W., Jr., Izett, G.A., 2005: Crystalline-rock ejecta and shocked minerals of the Chesapeake Bay impact structure, USGS-NASA Langley core, Hampton, Virginia with supplemental constraints on the age of impact (chapter E). *In*: Studies of the Chesapeake Bay impact structure-The USGS-NASA Langley corehole, Hampton, Vrigina and related coreholes ang geophysical surveys, edited by Horton, J.W., Jr., Powars, D.S. and Gohn, G.S., *U.S. Geological Survey Professional Paper* (1688), E1-E30.

Horton, W.J., Jr., Drake, A.A., Rankin, D.W., 1989: Tectonostratigraphic terranes and their Paleozoic boundaries in the central and southern Appalachians. *Geological Society of America Special Paper* (230), 213-245.

Horton, J.W., Jr., Kunk, M.J., Naeser, C.W., Naeser, N.D., Aleinkoff, J.N., Izett, G., 2002: Petrography, Geochronology, and significance of crystalline basement rocks and impact-derived clasts in the Chesapeake Bay impact structure, southeastern Virginia (abstract). *Geological Society of America Annual meeting*, 1 p.

Horton, W.J., Jr., Aleinikoff, J.N., Kunk, M.J., Naeser, C.W., Naeser, N.D., 2005a: Petrography, Stucture, age, and thermal history of granitic coastal plain basement in the Chesapeake Bay impact structure, USGS-NASA Langley core, Hampton, Virginia. *U.S. Geological Survey Special Paper* (1688), B1-B29.

Horton, J.W., Jr., Gohn, G.S., Jackson, J.C., Aleinikoff, J.N., Sanford, W.E., Edwards, L,E., Powars, D.S., 2005b: Results from a scientific test hole in the central uplift, Chesapeake Bay impact structure, Virginia, USA (abstract): *Lunar and Planetary Science Conference XXXVI*, 2 p.

Horton, J.W., Jr., Powars, D.S., Gohn, G.S. (editors), 2005c: Studies of the Chesapeake Bay impact structure-The USGS-NASA Langley corehole, Hampton, Virginia, and related coreholes and geophysical surveys. *U.S. Geological Survey Professional Paper* (1688), A1-K32.

Horton, J.W., Jr., Vank, D.A., Naeser, C.W., Naeser, N.D., Larsen, D., Jackson, J.C., Belkin, H.E., 2006: Postimpact hydrothermal conditions at the central uplift, Chesapeake Bay impact structure, Virginia, USA (abstract). *Lunar and Planetary Science Conference* (37), 1842.

References

Horton, J.W., Jr., Gohn, G.S., Powars, D.S., Edwards, L.E., 2008: Origin and emplacement of impactites in the Chesapeake Bay impact structure, Virginia, USA. *Geological Society of America Special Paper* (437), 73-98.

Horton, J.W., Jr., Gibson, R.L., Reimold, W.U., Wittmann, A., Gohn, G.S., Edwards, L.E., 2009a: Geologic columns for the ICDP-USGS Eyreville B core, Chesapeake Bay impact structure: Impactites and crystalline rocks, 1766 to 1096 m depth. *Geological Society of America Special Paper* (458), 21- 49.

Horton, J.W., Jr., Kunk, M.J., Belkin, H.E., Aleinikoff, J.N., Jackson, J.C., Chou, I-M., 2009b: Evolution of crystalline target rocks and impactites in the Chesapeake Bay impact structure, ICDP-USGS Eyreville B core. *Geological Society of America Special Paper* (458), 277-316.

Ivanov, B.A., 2001: Mars/Moon cratering rate ration estimates. *Space Science Review* (96), 87-104.

Izaola, Z., González, S., Elcoro, L., Perez-Mato, J.M., Madariaga, G., García, A., 2007: Revision of pyrrhotite structures within a common superspace model. *Acta Crystallographica* (B63), 693-702.

Jackson, M., 1991: Anisotropy of magnetic remanence: A brief review of mineralogical sources, physical origins, and geological applications, and comparison with susceptibility anisotropy. *Pure Applied Geophysics* (136), 1, 1-28.

Jackson, M., Moskowitz, B.M., Rosenbaum, J., Kissel, C., 1998: Field-dependence of AC susceptibility in titanomagnetites, *Earth and Planetary Science Letters* (157), 129-139.

Jackson, M., Moskowitz, B., Boweles, J., 2011: The magnetite Verwey transition. *IRM Quaterly* (20), 41, 7-11.

Janzen, M.P., Nicholson, R.V., Scharer, J.M., 2000: Pyrrhotite reaction kinetics: reaction rates for oxidation by oxygen, ferric iron, and for nonoxidative dissolution. *Geochimica et Cosmochimica Acta* (64), 9, 1511-1522.

Jensen, E., 1942: Pyrrhotite; melting relations and composition. *American Journal of Science* (240), 695-709.

Joreau, P., Leroux, H., Doukhan, J.-C., 1996: A transmission electron microscopy (TEM) investigation of opaque phases in shocked chondrites. *Meteoritcs and Planetary Science* (31), 305-312.

Jover, O., Rochette, P., Lorand, J.P., Maeder, M., Bouchez, J.L., 1989: Magnetic mineralogy of some granites from the French Massif Central: origin of their low-field susceptibility. *Physics of the Earth and Planetary Interiors* (55), 1-2, 79–92.

Just, J., Kontny, A., 2012: Thermally induced alterations of minerals during measurements of the temperature dependence of magnetic susceptibility: a case study from the hydrothermally altered Soultz-sous-Forêts granite, France. *International Journal of Earth Science* (101), 3, 819-839.

Kąkol, Z., Honig, J.M., 1989: Influence of deviations from ideal stoichiometry on the anisotropy parameters of magnetite $Fe_{3(1-\delta)}O_4$. *Physical Review B* (40), 13, 9090-9097.

Kamimura, T., Sato, M., Takahashi, H., Mori, N., Yoshida, H., Kaneko, T., 1992: Pressure-induced phase transition in Fe-Se and Fe-S systems with a NiAs-type structure. *Journal of Magnetism and Magnetic Materials* (104-107), 255-256.

Kenkmann, T., Schönian, F., 2006: Ries and Chicxulub: Impact craters on Earth provide insights for Martian ejecta blankets. *Meteoritics and Planetary Science* (41), 10, 1587-1603.

Kenkmann, T., Collins, G.S., Wittmann, A., Wünemann, K., Reimold, W.U., Melosh, H.J., 2009: A model for the formation of the Chesapeake Bay impact crater as revealed by drilling and numerical simulation. *Geological Society of America Special Paper* (458), 571-585.

Kieffer, S.W., Phakey, P.P., Christie, J.M., 1976: Shock processes in porous quartzite: Transmission electron microscope observations and theory. *Contributions to Mineralogy and Petrology* (59), 41-93.

Kissin, S.A., Scott, S.D., 1982: Phase relations involving pyrrhotite below 350°C. *Economic Geology* (77), 1739-1754.

Kitamura, M., Goto, T., Syono, Y., 1977: Intergrowth textures of diaplectic Glass and crystal in shock-loaded P-anorthite. *Contributions to Mineralogy and Petrology* (61), 299-304.

Kleeman, J.D., Ahrens T.J. 1973: Shock-induced transition of quartz to stishovite. *Journal of Geophysical Research* (78), 26, 5954-5960.

Kletetschka, G., Wasilewski, P.J., Taylor, P.T., 2000: Mineralogy of the sources for magnetic anomalies on Mars. *Meteoritics and Planetary Science* (35), 895-899.

Kobayashi, H., Sato, M., Kamimura, T., Sakai, M., Onodera, H., Kuroda, N., Yamaguchi, Y., 1997: The effect of pressure on the electronic states of FeS and Fe_7S_8 studied by Mössbauer spectroscopy. *Journal of Physics: Condensed Matter* (9), 515-527.

Kobayashi, K., 1959: Chemical remanent magnetization of ferromagnetic minerals and its application to rock magnetism. *Journal of Geomagnetism Geoelectrics* (10), 99-117.

Koeberl, C., 1994: Tektite origin by hypervelocity asteroidal or cometary impact: Target rocks, source craters, and mechanisms. *Geological Society of America Special Paper* (293), 133-151.

Koeberl, C., Poag, C.W., Reimold, W.U., Brandt, D., 1996: Impact origin of the Chesapeake Bay Structure and the source of the North American tektites. *Science* (271), 5253, 1263-1266.

Kohout, T., Kosterov, A., Haloda, J., Týcová, P., Zbořil, R., 2010: Low-temperature magnetic properties of iron-bearing sulfides and their contribution to magnetism of cometary bodies. *Icarus* (208), 955-962.

Kontny, A., De Wall, H., Sharp, T.G., Pósfai, M., 2000: Mineralogy and magnetic behavior of pyrrhotite from a 260°C section at the KTB drilling site, Germany. *American Mineralogist* (85), 1416-1427.

Kontny, A., Elbra ,T., Just, J., Pesonen, L.J., Schleicher, A.M., Zolk, J., 2007: Petrography and shock related remagnetization of pyrrhotite in drill cores from Bosumtwi impact crater drilling project, Ghana. *Meteoritics and Planetary Science* (42), 811-827.

Kosterov, A., 2001: Magnetic hysteresis of pseudo- single- domain and multidomain magnetite below the Verwey transition. *Earth and Planetary Science Letters* (186), 245-253.

Kosterov, A., 2003: Low-temperature magnetization and AC susceptibility of magnetite: Effect of thermomagnetic history. *Geophysical Journal International* (154), 58-71.

Krs, M., Novák, F., Pruner, P., Kouklíkavá, L., Jansa, J., 1992: Magnetic properties and metastability of greigite-smythite mineralization in brown-coal basins of the Krušné hory Piedmont, Bohemia. *Physics of the Earth and Planetary Interiors* (70), 273-287.

Kübler, L., 1985: Deformation mechanisms in experimentally deformed single crystals of pyrrhotite, $Fe_{1-x}S$. *Physics and Chemistry of MInerals* (12), 353-362.

Langenhorst, F., 1994: Shock experiments on pre-heated α- and β-quartz: II. X-ray and TEM investigations. *Earth and Planetary Science Letters* (128), 683-698.

Langenhorst, F., 2002: Shock metamorphism of some minerals: Basic introduction and microstructural observations, *Bulletin of the Czech Gelogical Survey* (77), 4, 265-282.

References

Langenhorst, F., Greshake, A., 1999: A transmission electron microscope study of Chassigny: Evidence for strong shock metamorphism. *Meteoritics and Planetary Science* (34), 43-48.

Langenhorst, F., Poirier, J.P., 2000: "Eclogitic" mineral in a shocked basaltic meteorite. *Earth and Planetary Science Letters* (176), 259-265.

Lattard, D., Engelmann, R., Kontny, A., Sauerzapf, U., 2006: Curie temperatures of synthetic titanomagnetites in the Fe-Ti-O system: Effects of composition, crystal chemistry, and thermomagnetic methods. *Journal of Geophysical Research* (111), doi:10.1029/2006JB004591.

Leroux, H., 2001: Microstructural shock signatures of major minerals in meteorites. *European Journal of Mineralogy* (13), 253-272.

Leroux, H., Reimold, W.U., Doukhan, J.-C., 1994: A TEM investigation of shock metamorphism in quartz from the Vredefort dome, South Africa. *Tectonophysics* (230), 223-239.

Li, F., Franzen, H.F., 1996: Ordering, incommensuration and phase transitions in pyrrhotite: Part II: A high-temperature X-ray powder diffraction and thermomagnetic study. *Journal of Solid State Chemistry* (126), 108-120.

Lian, S., Wang, E., Kang, Z., Bai, Y., Gao, L., Jiang, Y., Hu, C., Xu, L., 2004: Synthesis of magnetite nanorods and porous hematite nanorods. *Solid State Communications* (124), 485-490.

Linford, N., Linford, P., Platzman, E., 2005: Dating evironmental change using magnetic bacteria in archaeological soils from the upper Thames Valley, UK. *Journal of Archaeological Science* (32), 1037-1043.

Liu, Q., Banjaree, S.K., Jackson, J.M., 2003: An integrated study of the grain-size-dependent magnetic mineralogy of the Chinese loess/paleosol and its environmental significance. *Journal of Geophysical Research* (108), B9, 2437, doi: 10.1029/2002JB002264.

Louzada, K., 2008: Imaging of experimentally shocked pyrrhotite. *IRM Quaterly* (18), 2, 4-5.

Louzada, K. L., Stewart, S. T., Weiss, B.P., 2007: Effect of shock on the magnetic properties of pyrrhotite, the Martian crust, and meteorites. *Geophysical Research Letters* (34), L05204, doi: 10.1029/2006GL027685.

Louzada, K. L., Stewart, S. T., Weiss, B. P., Gattacceca, J., Bezaeva, N. S., 2010: Shock and static pressure demagnetization of pyrrhotite and implications for the Martian crust. *Earth and Planetary Science Letters* (290), 90-101.

Louzada, K. L., Stewart S. T., Weiss B. P., Gattacceca, J., Lillis R. J., Halekas, J. S., 2011: Impact demagnetization of the Martian crust: Current knowledge and future directions. *Earth and Planetary Science Letters* (305), 257-269.

Lovlie, R., Opdyke, N.D., 1974: Rock magnetism and paleomagnetism of some intrusions from Virginia. *Journal of Geophysical Research* (79), 343-349.

Mang, C., Harries, D. Kontny, A., Langenhorst, F., Hecht, L., 2012: Shock deformation of pyrrhotite in the suevites of the Chesapeake Bay impact crater, USA. *Meteoritics and Planetary Science Letters* (47), 2, 277-295.

Mann, S., Sparks, N.H.C., Blakemore, R.P., 1987: Structure, morphology and crytsal growth of anisotropic magnetite crystals in magnetotactic bacteria. *Proceedings of the Royal Society of London. Series B, Biological Sciences* (231), 1265, 477-487.

Marshall, R.R., 1961: Devitrification of natural glass. *Geological Society of America Bulletin* (72), 1493-1520.

Matsunaga, T., Sakaguchi, T., Tadokoro, F., 1991: Magnetite formation by a magnetic bacterium. *Applied Microbiology and Biotechnology* (35), 551-655.

McCall, G.H.J., 2009: Half a century of progress in research on terrestrial impact structures: A review. *Earth-Science Reviews* (92), 3-4, 99-116.

Melosh, H.J., 1989: Impact cratering - a geologic process. *Oxford University Press*, New York, 253 p.

Melosh, H.J., Ivanov, B.A., 1999: Impact crater collapse. *Annual Reviews Earth and Planetary Science* (27), 385-415.

Meyer, C., Fritz, J., Misgaiski, M., Stöffler, D., Artemieva, N.A., Hornemann, U., Moeller, R., De Vera, J.-P., Cockell, C., Horneck, G., Ott, S., Rabbow, E., 2011: Shock experiments in support of the Lithopanspermia theory: The influence of host rock composition, temperature, and shock pressure on the survival rate of endolithic and epilithic microorganisms. *Meteoritics and Planetary Science* (46), 5, 701-718.

Michel, A., Chaudron, G., Bénard, J., 1951: Propriétés des composes ferromagnétiques non méttaliques. *Journal de Physique et le Radium* (12), 3, 189-201.

Morimoto, N., Gyobu, A., Tsukuma, K., Koto, K., 1975: Superstructure and nonstoichiometry of intermediate pyrrhotite. *American Mineralogist* (60), 240-248.

Moskowitz, B.M., 1981: Methods for estimating Curie temperatures of titanomagnetites from experimental Js-T data. *Earth and Planetary Science Letters* (53), 84-88.

Moskowitz, B.M., Frankel, R.B., Flanders, P.J., Blakemore, R.P., Schwartz, B.B., 1988: Magnetic properties of magnetotactic bacteria. *Journal of Magnetism and Magnetic Materials* (73), 273-288.

Moskowitz, B.M., Frankel, R.B., Bazylinski, D.A., Jannasch, H.W., Lovley, D.R., 1989: A comparison of magnetite particles produced anerobically by magnetotactic and dissimilatory iron-reducing bacteria. *Geophysical Research Letters* (16), 7, 665-668.

Moskowitz, B.M., Jackson, M., Kissel, C., 1998: Low-temperature magnetic behavior of titanomagnetites. *Earth and Planetary Science Letters* (157), 141-149.

Müller, W.F., Hornemann, U., 1969: Shock-induced planar deformation structures in experimentally shock-loaded olivines and in olivines from chondritic meteorites. *Earth and Planetary Science Letters* (7), 3, 251-264.

Murowchick, J.B., 1992: Marcasite inversion and the petrographic determination of pyrite ancestry. *Economic Geology* (87), 1141-1152.

Muxworthy, A.R., Dunlop, D.J., 2002: First- order reversal curve (FORC) diagrams for pseudo-single-domain magnetites at high temperature. *Earth and Planetary Science Letters* (203), 369-382.

Muxworthy, A.R., Mc Clelland, E., 2000: Review of the low-temperature magnetic properties of magnetite from a rock magnetic perspective. *Geophysical Journal International* (140), 101-114.

Muxworthy, A.R., Dunlop, D.J., Özdemir, Ö., 2003: Low-temperature cycling of isothermal and anhysteretic remanence: microcoercivity and magnetic memory. *Earth and Planetary Science Letters* (205), 173-184.

Nagata, T., 1971: Introductory notes on shock remanent magnetization and shock demagnetizahion of igneous rocks. *Pure and Applied Geophysics* (89), 1, 159-177.

Nakazawa, H., Morimoto, N., 1971: Phase relations and superstructures of pyrrhotite, $Fe_{1-x}S$. *Materials Research Bulletin* (6), 5, 345-357.

Neff, D., Dillmann, P., Bellot-Gurlet, L., Beranger, G., 2005: Corrosion of iron archeological artefacts. *Corrosion Science*

(47), 515-535.

Newsom, H.E., Graup, G., Sewards, T., Keil, K., 1986: Fluidization and hydrothermal alteration of the suevite deposit at the Ries crater, West Germany and implications for Mars. *Journal of Geophysical Research* (91), B13, B239-E251.

Niederschlag, E., Siemes, H., 1996: Influence of initial texture, temperature and total strain on the texture development of polycrystalline pyrrhotite ores in deformation experiments. *Textures and Microstructures* (28), 129-148.

O'Reilly, W., 1984: Rock and mineral magnetism. *Blackie*, Glasgow and New York, 220 p.

Ormö, J., Lindström, M., 2000: When a cosmic impact strikes the sea bed. *Geological Magazine* (137), 67-80.

Osinski, G.R., Grieve, R.A.F., Spray, J.G., 2004: The nature of the groundmass of surficial suevite from the Ries impact structure, Germany, and constraints on its origin. *Meteoritics and Planetary Science* (39), 10, 1655-1683.

Özdemir, Ö., Banerjee, S.K., 1984: High temperature stability of maghemite (γ-Fe_2O_3). *Geophysical Research Letters* (11), 3, 161-164.

Özdemir, Ö., Dunlop, D.J., 1993: Chemical remanent magnetization during γ-FeOOH transformations. *Journal of Geophysical Research* (98), B3, 4191-4198.

Özdemir, Ö., Dunlop, D.J., 1999: Low-temperature properties of a single crystal of magnetite oriented along principal magnetic axes. *Earth and Planetary Science Letters* (165), 2, 229-239.

Özdemir, Ö., Dunlop, D. J., 2003: Low-temperature behavior and memory of iron-rich titanomagnetites (Mt. Haruna, Japan and Mt. Pinatubo, Philippines). *Earth and Planetary Science Letters* (216), 193-200.

Özdemir, Ö., Dunlop, D.J., Moskowitz, B.M., 1993: The effect of oxidation on the Verwey transition in magnetite. *Geophysical Research Letters* (20), 16, 1671-1674.

Özdemir, Ö., Dunlop, D.J., Moskowitz, B.M., 2002: Changes in remanence, coercivity and domain state at low temperature in magnetite. *Earth and Planetary Science Letters* (194), 343-358.

Pardoe, H., Chua-Anusorn, W., St. Pierre, T.G., Dobson, J., 2001: Structural and magnetic properties of nanoscale iron oxide particles synthesized in the presence of dextran or polyvinyl alcohol. *Journal of Magnetism and Magnetic Materials* (225), 41-46.

Passchier, C.W., Trouw, R.A.J., 2005: Microtectonics. *Springer*, Heidelberg, 366 p.

Penn, R.L., Erbs, J.J., Gulliver, D.M., 2006: Controlled growth of aplpha-FeOOH nanorods by exploiting-oriented aggregation. *Journal of Crystal Growth* (293), 1-4.

Petrík, I., Broska, I., 2007: Petrology of two granite types from the Tribeč Mountains, Western Carpatians: An example of allanite (+magnetite) versus monazite dichotomy. *Geological Journal* (29), 1, 59-78.

Petrovský, E., Kapička, A., 2006: On the determination of the Curie point from thermomagnetic curves. *Journal of Geophysical Research* (111), B12S27, doi: 10.1029/2006JB004507.

Pilkington, M., Grieve, R.A.F., 1992: The geophysical signature of terrestrial impact craters. *Review of Geophysics* (30), 161-181.

Plado, J., Pesonen, L.J., Koeberl, C., Elo, S., 2000: The Bosumtwi meteorite impact structure, Ghana: A magnetic model. *Meteoritics and Planetary Science* (35), 723-732.

References

Pleiscia, J.B., Daniels, D.I., Sha, A.K., 2009: Gravity investigations of the Chesapeake Bay impact structure, *Geological Society of America Special Paper* (458), 181-194.

Pohl, J., Angenheister, G., 1969: Anomalien des Erdmagnetfeldes und Magnetisierung der Gesteine im Nördlinger Ries. *Geologica Bavarica* (61), 327-336.

Pohl, J., Poschlod, K., Reimold, W.U, Meyer, C., Jacob, J., 2010: Ries crater, Germany: The Enkingen magnetic anomaly and associated drill core SUBO 18. *Geological Society of America Special Paper* (465), 141-163.

Pósfai, M., Sharp, T. G., Kontny, A., 2000: Pyrrhotite varieties from the 9.1 km deep borehole of the KTB project. *American Mineralogist* (85), 1406-1415.

Powars, D.S., Bruce, T.S., 1999: The effects of the Chesapeake Bay impact crater on the geological framework and correlation of hydrogeologic units of the lower York-James Peninsula, Virginia. U.S. *Geolological Survey Professional Paper* (1612), 82 p.

Powars, D.S, Catchings, R.D., Goldman, M.R., Gohn, G.S., Horton, J.W., Jr., Edwards, L.E., Rymer, M.J., Gandhok, G., 2009: High-resolution seismic-reflection images across the ICDP-USGS Eyreville deep drilling site, Chesapeake Bay impact structure, *Geological Society of America Special Paper* (458), 209-233.

Pratt, A.R., Muir, I.J., Nesbitt, H.W., 1994: X-ray photoelectron and Auger electron spectroscopic studies of pyrrhotite and mechanism of air oxidation. *Geochimica et Cosmochimica Acta* (58), 827-841.

Putnis, A., 1975: Observations on coexisting pyrrhotite phases by transmission electron microscopy. *Contributions to Mineralogy and Petrology* (52), 307-313.

Putnis, A., 1992: Introduction to mineral sciences. *Cambridge University Press*, Cambridge, 458 p.

Ramasesha, S.K., Mohan, M., Singh, A.K., Honig, J.M., Rao, C.N R., 1994: High-pressure study of Fe_3O_4 through the Verwey transition. *Physical Review B* (50), 18, 789-791.

Reimold, W.U., Koeberl, C., 2008: Catastrophes, extinctions and evolution: 50 years of impact cratering studies. *Geological Society of India* (66), 69-110.

Reimold, W.U., Hansen, B.K., Jacob, J., Artemievau, N.A., Wünnemann, K., Meyer, C., 2012: Petrography of the impact breccias of the Enkingen (SUBO 18) drill core, southern Ries crater, Germany: New estimate of impact melt volume. *Geological Society of America Bulletin* (124), 1/2, 104-132.

Ressetar, R., Martin, D.L., 1980: Paleomagnetism of Eocene Igneous intrusions in the Valley and Ridge Province, Virginia and West Virginia. *Canadian Journal of Earth Sciences* (17), 1, 583-588.

Roberts, A.P., Pike, C.R., Verosub, K.L., 2000: First-order reversal curve diagrams: A new tool for characterizing the magnetic properties of natural samples. *Journal of Geophysical Research* (105), B12, 28,461-28,475.

Rochette, P., 1987: Metamorphic control of the magnetic mineralogy of black shales. *Earth and Planetary Science Letters* (84), 446-456.

Rochette, P., Fillion, G., Mattéi, J.-L. and Dekkers, M.J., 1990: Magnetic transition at 30-34 Kelvin in pyrrhotite: insight into a widespread occurrence of this mineral rocks. *Earth and Planetary Science Letters* (98), 319- 328.

Rochette, P., Dekkers, M.J., van Velzen, A.J., Hoffmann, V., Horng, C.S., 1994: Magnetic properties of various natural sulfides at high and low temperature. *Annales de Géophysique* (supplement), (12), C118.

References

Rochette, P., Lorand, J.-P., Fillion, G. and Sautter, V., 2001: Pyrrhotite and the remanent magnetization of SNC meteorites: a changing perspective on Martian magnetism. *Earth and Planetary Science Letters* (190), 1-12.

Rochette, P., Fillion, G., Ballou, R., Brunet, F., Ouladdiaf, B., Hood, L., 2003: High pressure magnetic transition in pyrrhotite and impact demagnetization on Mars. *Geophysical Research Letters* (30), 13, 1683, doi: 10.1029/2003GL017359.

Rochette, P., Gattacceca, J., Chevrier, V., Hoffmann, V., Lorand, J.-P., Funaki, M., Hochleitner, R., 2005: Matching Martian crustal magnetization and magnetic properties of martian meteorites. *Meteoritics and Planetary Science* (40), 4, 529-540.

Roggwiller, P., Kündig, W., 1973: Mössbauer spectra of superparamagnetic Fe_3O_4. *Solid State Communications* (12), 901-903.

Schmitt, R.T., 2000: Shock experiments with the H6 chondrite Kernouvé: Pressure calibration of microscopic shock effects. *Meteoritics and Planetary Science* (35), 545-560.

Schwarz, E.J., Vaughan, D.J., 1972: Magnetic phase relations of pyrrhotite, *Journal of Geomagnetism and Geoelectricity* (24), 441-458.

Shah, A.K., Brozena, J., Vogt, P., Daniles, D., Plescia, J., 2005: New surveys of the Chesapeake Bay impact structure suggest melt pockets and target structure effect. *Geology* (33), 417-420.

Shah, A.K., Daniels, D.L., Kontny, A., Brozena, J., 2009: Megablocks and melt pockets in the Chesapeake Bay impact structure constrained by magnetic field measurements and properties of the Eyreville and Cape Charles cores. *Geological Society of America Special Paper* (458), 195-209.

Sharp, T.G., Lingemann, C.M., Dupas, C., Stöffler, D., 1997: Natural occurence of $MgSiO_3$-ilmenite and evidence for $MgSiO_3$-perovskite in a shocked L-chondrite. *Science* (277), 352-355.

Shepherd, J.P., Aragón, R., Koenitzer, J.W., Honig, J.M., 1985: Changes in the nature of the Verwey transition in nonstoichiometric magnetite (Fe_3O_4). *Physical Review B* (32), 3, 1818-1819.

Shoemaker, E.M., 1977: Why study impact craters? *In*: Impact and explosion cratering: Planetary and terrestrial implications, *Pergamon Press, Inc.*, New York, 1-10.

Soffel, H.C., 1991: Paläomagnetismus und Archäomagnetismus. *Springer*, Heidelberg, 276 p.

Stähle, V., 1972: Impact glasses from suevite of the Nördlinger Ries. *Earth and Planetary Science Letters* (17), 275-293.

Stöffler, D., 1966: Zones of impact metamorphism in the crystalline rocks of the Nördlinger Ries crater. *Contributions to Mineralogy and Petrology* (12), 1, 15-24.

Stöffler, D., 1971: Progressive metamorphism and classification of shocked and brecciated crystalline rocks at impact craters. *Journal of Geophysical Research* (76), 23, 5541-5551.

Stöffler, D., Grieve, R.A.F., 1994: Classification and Nomenclature of impact metamorphic rocks: A proposal to the IUGS Subcomission on the systematics of metamorphic rocks (abstract). *Lunar and Planetary Science Conference* (25), 1347.

Stöffler, D., Langenhorst, F., 1994: Shock metamorphism of quartz in nature and experiment: I. Basic observation and theory. *Meteoritics* (29), 155-181.

Stöffler, D., Ostertag, R., Jammes, C., Pfannschmidt, G., 1986: Shock metamorphism and petrography of the Shergotty achondrite. *Geochimica et Cosmochimica Acta* (50), 889-903.

References

Stöffler, D., Keil, K., Scott, E.R.D., 1991: Shock metamorphism of ordinary chondrites. *Geochimica et Cosmochimica Acta* (55), 3845-3867.

Stöffler, D., Ryder, G., Ivanov, B.A., Artemieva, N.A., Cintala, M.J., Grieve, R.A.F., 2006: Cratering history and lunar chronology. *Reviews in Mineralogy and Geochemistry* (60), doi: 10.2138/rmg.2006.60.05.

Taylor, L.A., Williams, K.L., 1972: Smythite, $(Fe, Ni)_9S_{11}$ A redefinition. *American Mineralogist* (57), 1571-1577.

Therriault, A.M., Fowler, A.D., Grieve, R.A.F., 2002: The Sudbury igneous complex: A differentiated impact melt sheet. *Economic Geology* (97), 7, 1521-1540.

Todo, S., Takeshita, N., Kanehara, T., Mori, T., Mori, N., 2001: Metallization of magnetite (Fe_3O_4) under high pressure. *Journal of Applied Physics* (89), 7347, doi: 10.1063/1.1359460.

Townsend, N.T., Gibson, R.L., Horton, J.W., Jr., Reimold, W.U., Schmitt, R.T., Bartosova, K., 2009: Petrographic and geochemical comparisons between the lower crystalline basement-derived section and the granite megablock and amphibolite megablock of the Eyreville B core, Chesapeake Bay impact structure, USA. *Geological Society of America Special Paper* (458), 255-275.

Ugalde, H., Morris, W.A., Pesonen, L.J., Danuor, S.K., 2007: The Lake Bosumtwi meteorite impact structure, Ghana: Where is the magnetic source? *Meteoritics and Planetary Science* (42), 4-5, 867–882.

Urrutia-Fucugauchi, J., Soler-Arechalde, A.M., Renolledo-Vieyra, M., Vera-Sanchez, P., 2004: Paleomagnetic and rock magnetic study of the Yaxcopoil-1 impact breccia sequence, Chixculub impact crater (Mexico). *Meteoritics and Planetary Science* (39), 6, 843-856.

Usselmann, T.M., 1975a: Experimental approach to the state of the core: Part I. The Liquids relations of the Fe-rich portion of the Fe-Ni-S-system from 30 to 100 kb. *American Journal of Science* (275), 278-290.

Usselmann, T.M., 1975b: Experimental approach to the state of the core: Part II. Composition and thermal regime. *American Journal of Science* (275), 291-303.

Vahle, C., Kontny, A., Gunnlaugsson, H.P., Kristjansson, L., 2007: The Stardalur magnetic anomaly revisited-new insights into a complex cooling and alteration history. *Physics of the Earth Planetary Interiors* (164), 119-141.

Vandenberghe, R.E., Grave, E., Bakker, P.M.A., Krs, M., 1991: Mössbauer effect study of natural greigite. *Hyperfine Interactions* (68), 319-322.

Van Landuyt, J., Amelinckx, S., 1972: Electron microscope observations of the defect structure of pyrrhotite. *Materials Research Bulletin* (7), 71-79.

Vaughan, D.J., Schwarz, E.J., Owens, D.R., 1971: Pyrrhotites from the Strathcona Mine, Sudbury, Canada; A thermomagnetic and mineralogical study. *Economic Geology* (66), 1131-1144.

Vernaz, E., Gin, S., Jégou, C., Ribet, I., 2001: Present understanding of R7T7 glass alteration kinetics and their impact on long-term behavior modeling. *Journal of Nuclear Materials* (298), 27-36.

Verwey, E.J.W., 1939: Electronic conduction of magnetite (Fe_3O_4) and its transition point at low temperatures. *Nature* (144), 327-328.

Wang, G., Whittaker, G., Harrison, A., Song, L., 1998: Preparation and mechanism of formation of goethite-magnetite particles by decomposition of ferric and ferrous salts in aqueous solution using microwave radiation. *Materials Research*

References

Bulletin (33), 11, 1571-1579.

Watmuff, I.G., 1974: Supergene alteration of the Mt. Windarra nickel sulphide ore deposit, Western Australia. *Mineralium Deposita* (9), 199-221.

Weiss, B.P., Kim, S.S., Kirschvink, J.L., Kopp, R.E, Sankaran, M., Kobayashi, A., Komeili, A., 2004: Magnetic tests for magnetosome chains in Martian meteorite ALH84001. *PNAS* (101), 22, 8281-8284.

Williams, H.J., Bozorth, R.M., Goertz, M., 1953: Mechanism of transition in magnetite at low temperatures. *Physical Review* (91), 5, 1107-1115.

Williams, W., Dunlop, D.J., 1989: Three-dimensional micromagnetic modelling of ferromagnetic domain structure. *Nature* (337), 634-637.

Williams, Q., Jeanloz, R., 1989: Static amorphization of anorthite at 300 K and comparison with diaplectic glass. *Nature* (338), 413-415.

Wittmann, A., Reimold, W.U., Schmitt, R.T., Hecht, L., Kenkmann, T., 2009a: The record of ground zero in the Chesapeake Bay impact crater-suevites and related rocks. *Geological Society of America Special Paper* (458), 349-376.

Wittmann, A., Schmitt, R.T., Hecht, L., Kring, D.A., Reimold, W.U., Povenmire, H., 2009b: Petrology of impact melt rocks from the Chesapeake Bay crater, USA. *Geological Society of America Special Paper* (458), 377-396.

Wolfers, P., Fillion, G., Ouladdiaf, B., Ballou, R., Rochette, P., 2011: The Pyrrhotite 32K magnetic transition, *Solid State Phenomena* (170), 174-179.

Wünnemann, K., Collins, G.S., Melosh, H.J., 2006: A strain-based porosity model for use in hydrocode simulations of impacts and implications for transient crater growth in porous targets. *Icarus* (180), 2, 514-527.

Zapletal, K., 1993: Effect of intergrowths of the ferromagnetic and antiferromagnetic phases on the rock magnetic properties of natural pyrrhotites, *Physics of the Earth and Planetary Interiors* (76), 151-162.

Abbreviations

AF: alternating field
B: Bayside
CC: Cape Charles
χ: magnetic susceptibility
χ': magnetic in-phase susceptibility
χ'': magnetic out-of-phase susceptibility
DC: direct current
DFG: Deutsche Forschungsgesellschaft
Eyr: Eyreville
FC: field cooled
FIB: focused iron beam
Fig.: figure
H_C: coercivity
H_{cr}: coercivity of remanence
hem: hematite
hex: hexagonal
incl.: inclination
M_{rs}: saturation remanent magnetization
M_s, J_s: saturation magnetization
HT: high temperature
ilm: ilmenite
IRM: isothermal remanent magnetization
J_i: induced magnetization
KIT: Karlsruher Institut für Technologie
L: Langley
LT: low temperature
MD: multidomain
MDF: mean destructive field
mon: monoclinic
mrc: marcasite
mt: magnetite
NRM: natural remanent magnetization
n: number
OM: optical microscope
p: pressure
PDF: planar deformation features
PF: planar fractures
po: pyrrhotite
prim: primary
PSD: pseudo-single-domain
py: pyrite
Q-ratio: Königsberger factor
RT: room temperature
SAED: selected area electron diffraction

SD: single domain
sec: secondary
SEM: scanning electron microscope
shd: shocked
SIRM: saturation isothermal remanent magnetization
smy: smythite
sph: sphene
st.dev.: standard deviation
suev: suevite
T_B: blocking temperature
T_C: Curie temperature
TEM: transmissions electron microscope
therm: thermal
T_V: Verwey transition temperature
USGS: United States Geological Survey
ZFC: Zero field cooled

i want morebooks!

Buy your books fast and straightforward online - at one of world's fastest growing online book stores! Environmentally sound due to Print-on-Demand technologies.

Buy your books online at
www.get-morebooks.com

Kaufen Sie Ihre Bücher schnell und unkompliziert online – auf einer der am schnellsten wachsenden Buchhandelsplattformen weltweit! Dank Print-On-Demand umwelt- und ressourcenschonend produziert.

Bücher schneller online kaufen
www.morebooks.de

 VDM Verlagsservicegesellschaft mbH
Heinrich-Böcking-Str. 6-8   Telefon: +49 681 3720 174   info@vdm-vsg.de
D - 66121 Saarbrücken   Telefax: +49 681 3720 1749   www.vdm-vsg.de

Printed by Books on Demand GmbH, Norderstedt / Germany